A Science Handbook for Musicians, Entrepreneurs, and Candidates for Public Office

A Science Handbook for Musicians, Entrepreneurs, and Candidates for Public Office

Donald Greenspan, Ph.D.

VANTAGE PRESS
New York

Illustration herein by Gary Oliver, used with permission.

FIRST EDITION

Published by Vantage Press, Inc.
516 West 34th Street, New York, New York 10001

Manufactured in the United States of America
ISBN: 0-533-14163-X

Library of Congress Catalog Card No.: 01-130495

0 9 8 7 6 5 4 3 2 1

Contents

Plates

A Science Handbook
for
Musicians, Entrepreneurs,
and Candidates for Public Office

Prologue

U.S. Report Fears Most Americans Will Become Scientific Illiterates

—*The New York Times*, Thursday, October 23, 1980

This handbook is founded on the observation that the larger concepts in science are often simple to understand, while the details may be beyond the reach of all but a few professionals. This notion was impressed upon me by Richard Feynman.

The book is divided into two parts. The first part gives the reader an overview of major scientific results, which should enable him or her to form reasonable decisions about related political or social issues. The second part is a dictionary of scientific terms, which can be consulted when reading articles of a scientific nature in newspapers and magazines.

Part One

The Sciences

I

Checks and Balances in Scientific Discovery

Science is the study of Nature. It is all-inclusive and examines such diverse matters as reproduction, energy, magnets, tidal waves, volcanic activity, X rays, food production, plant and human diseases, the motions of planets, and the evolution of stars. The amount of scientific knowledge available is so large that it is categorized formally into named disciplines, the basic ones being biology, chemistry, and physics. These three are themselves so large that they are often subdivided into specialized areas like zoology, spectroscopy, geology, metallurgy, genetics, and astronomy. Also, there are numerous areas in which different sciences overlap.

In order to develop a broad overview, we will restrict attention to vital aspects of the three basic sciences, biology, chemistry, and physics, and to two specialized sciences, astronomy and genetics. In this fashion we will be able to discuss topics of contemporary interest and importance, like cloning, nuclear energy, and human health.

One of the main reasons why a scientist studies Nature is that he or she would like to control its hidden forces. Understanding the ways in which Nature works might enable us to grow more food, to prevent normal cells from becoming malignant, to develop inexpensive sources of energy, and to prevent environmental pollution. In cases in which control may not be possible, like earthquake activity or volcanic eruption, the scientist would like to be able to predict the time of such events, thus providing an early warning system and preventing excessive loss of life.

The discovery of knowledge by scientific means is carried out in a unique way. First, a group of experimental scientists, working as meticulously as possible, reach conclusions from experiments and observations. Since no one is perfect, not even a scientist, all experimental conclusions have some degree of error. Hopefully, the error is small. Then there are the theoretical scientists, who concoct theories from which conclusions are reached, using strict mathematical reasoning from a set of assumptions, often called axioms. Experimental scientists are constantly checking theoretical conclusions by designing and carrying out new experiments. Theoretical scientists are constantly refining their theories by incorporating new experimental results into their axioms. The two groups thus work in a constant check and balance refinement process to create knowledge. And only after extensive experimental verification and widespread professional agreement is a scientific conclusion accepted as valid. The usual rule for acceptance is that the knowledge created is valid at least 95 percent of the time, from which follows the allowance of a maximum of 5 percent error in scientific work.

It should be clear that all scientific knowledge, because of the way it is derived and because human error may be involved, is never perfect. Scientists must live with knowing that they never know anything exactly. A belief that one knows something exactly is a religious belief. However, scientific knowledge may be of such high quality that its value is undeniable to most people. For example, after worldwide testing, it was concluded that the Salk and Sabin vaccines prevent polio, because the vaccines worked nearly 100 percent of the time. There are, however, a few isolated cases of children in whom the vaccines have had negative effects. Nevertheless, the vaccines' value in the improvement of world health is in no way diminished.

Thus, science is not perfect. However, history has proved it to be the best known and most reliable means of understanding the world around us. Its achievements have been remarkable, and many of these will be discussed in the chapters that follow.

It might be indicated, but not elaborated upon here, that mathematics is not a science, nor is engineering. Mathematics provides the analytical tools necessary for doing scientific research, while engineering provides the physical tools for utilizing the results of scientific research.

II

Biology—the Study of All Living Things

The Fundamental Natural Balance

All living things, both plants and animals, thrive in a simple but beautiful symbiotic natural balance. Animals require a certain gas called oxygen to derive energy from food. Plants require a certain gas called carbon dioxide to convert sunlight into food. Amazingly enough, animals inhale oxygen from the air and exhale carbon dioxide, while green plants "inhale" carbon dioxide and "exhale" oxygen.

To explain this phenomenon, biologists assume that the sun is our major source of energy. The Sun's energy is converted into food by the green portion of plants, like leaves, by a process called photosynthesis. Plants and plant products, like fruit and vegetables, are consumed and support the life processes of plant eating (herbivorous) animals, like rabbits, sheep, cows and deer. In turn, herbivorous animals are consumed by and support the life processes of the meat-eating (carnivorous) animals, like eagles, lions, and sharks. When sufficiently hungry, many animals can be both herbivorous and carnivorous.

In order for an animal to be able to use its food for energy and body repair, it needs to have a special burning agent for the food. This agent is a gas called oxygen, which is available abundantly in both air and water. Animals that live in air breathe oxygen with their lungs. Fish have specially adapted organs called gills for the extraction of oxygen from water. Seals, porpoises, and whales, which live in water but are not fish, have lungs, not gills, and must surface to inhale their oxygen from the air.

After food has been used within an animal's body, there are refuse products, which must be eliminated. One of these is called carbon, and it combines with the oxygen to form a gas called carbon dioxide. Carbon dioxide is eliminated from the lungs when one exhales. So, animals are constantly inhaling oxygen and exhaling carbon dioxide.

Now, if the atmosphere and the oceans were not constantly replenished with oxygen, all animal life might soon die. Nature now performs a magnificent balancing act, for the green portion of plants requires carbon dioxide to make food and as a byproduct emits oxygen. In this way all living things, the plants and animals, coexist in a mutually beneficial fashion.

Individuals who are unaware of this natural balance may be prone to cutting down trees mindlessly.

Dietary Supplements

The food intake of each animal must contain certain basic nutrients for the preservation of the health of that animal. For the human animal, these nutrients include vitamins and minerals, and both are available abundantly from fresh fruit and vegetables when eaten in variety. Also, because of the scientific understanding of the composition of vitamins and minerals, they are now produced synthetically and are available in pill or liquid form in many parts of the world. Claims about herbs usually do not have scientific

FOOD AND NUTRITION BOARD, NATIONAL ACADEMY OF SCIENCES—NATIONAL RESEARCH COUNCIL
RECOMMENDED DIETARY ALLOWANCES,[a] Revised 1989

Designed for the maintenance of good nutrition of practically all healthy people in the United States

							Fat-Soluble Vitamins				Water-Soluble Vitamins							Minerals						
Category	Age (years) or Condition	Weight[b] (kg)	Weight[b] (lb)	Height[b] (cm)	Height[b] (in)	Protein (g)	Vitamin A (μg RE)[c]	Vitamin D (μg)[d]	Vitamin E (mg α-TE)[e]	Vitamin K (μg)	Vitamin C (mg)	Thiamin (mg)	Riboflavin (mg)	Niacin (mg NE)[f]	Vitamin B_6 (mg)	Folate (μg)	Vitamin B_{12} (μg)	Calcium (mg)	Phosphorus (mg)	Magnesium (mg)	Iron (mg)	Zinc (mg)	Iodine (μg)	Selenium (μg)
Infants	0.0–0.5	6	13	60	24	13	375	7.5	3	5	30	0.3	0.4	5	0.3	25	0.3	400	300	40	6	5	40	10
	0.5–1.0	9	20	71	28	14	375	10	4	10	35	0.4	0.5	6	0.6	35	0.5	600	500	60	10	5	50	15
Children	1–3	13	29	90	35	16	400	10	6	15	40	0.7	0.8	9	1.0	50	0.7	800	800	80	10	10	70	20
	4–6	20	44	112	44	24	500	10	7	20	45	0.9	1.1	12	1.1	75	1.0	800	800	120	10	10	90	20
	7–10	28	62	132	52	28	700	10	7	30	45	1.0	1.2	13	1.4	100	1.4	800	800	170	10	10	120	30
Males	11–14	45	99	157	62	45	1,000	10	10	45	50	1.3	1.5	17	1.7	150	2.0	1,200	1,200	270	12	15	150	40
	15–18	66	145	176	69	59	1,000	10	10	65	60	1.5	1.8	20	2.0	200	2.0	1,200	1,200	400	12	15	150	50
	19–24	72	160	177	70	58	1,000	10	10	70	60	1.5	1.7	19	2.0	200	2.0	1,200	1,200	350	10	15	150	70
	25–50	79	174	176	70	63	1,000	5	10	80	60	1.5	1.7	19	2.0	200	2.0	800	800	350	10	15	150	70
	51+	77	170	173	68	63	1,000	5	10	80	60	1.2	1.4	15	2.0	200	2.0	800	800	350	10	15	150	70
Females	11–14	46	101	157	62	46	800	10	8	45	50	1.1	1.3	15	1.4	150	2.0	1,200	1,200	280	15	12	150	45
	15–18	55	120	163	64	44	800	10	8	55	60	1.1	1.3	15	1.5	180	2.0	1,200	1,200	300	15	12	150	50
	19–24	58	128	164	65	46	800	10	8	60	60	1.1	1.3	15	1.6	180	2.0	1,200	1,200	280	15	12	150	55
	25–50	63	138	163	64	50	800	5	8	65	60	1.1	1.3	15	1.6	180	2.0	800	800	280	15	12	150	55
	51+	65	143	160	63	50	800	5	8	65	60	1.0	1.2	13	1.6	180	2.0	800	800	280	10	12	150	55
Pregnant						60	800	10	10	65	70	1.5	1.6	17	2.2	400	2.2	1,200	1,200	320	30	15	175	65
Lactating	1st 6 months					65	1,300	10	12	65	95	1.6	1.8	20	2.1	280	2.6	1,200	1,200	355	15	19	200	75
	2nd 6 months					62	1,200	10	11	65	90	1.6	1.7	20	2.1	260	2.6	1,200	1,200	340	15	16	200	75

[a] The allowances, expressed as average daily intakes over time, are intended to provide for individual variations among most normal persons as they live in the United States under usual environmental stresses. Diets should be based on a variety of common foods in order to provide other nutrients for which human requirements have been less well defined. See text for detailed discussion of allowances and of nutrients not tabulated.

[b] Weights and heights of Reference Adults are actual medians for the U.S. population of the designated age, as reported by NHANES II. The median weights and heights of those under 19 years of age were taken from Hamill et al. (1979) (see pages 16–17). The use of these figures does not imply that the height-to-weight ratios are ideal.

[c] Retinol equivalents. 1 retinol equivalent = 1 μg retinol or 6 μg β-carotene. See text for calculation of vitamin A activity of diets as retinol equivalents.

[d] As cholecalciferol. 10 μg cholecalciferol = 400 IU of vitamin D.

[e] α-Tocopherol equivalents. 1 mg d-α tocopherol = 1 α-TE. See text for variation in allowances and calculation of vitamin E activity of the diet as α-tocopherol equivalents.

[f] 1 NE (niacin equivalent) is equal to 1 mg of niacin or 60 mg of dietary tryptophan.

Plate I. National Academy of Sciences Recommended Dietary Allowances For Good Nutrition.

support, while the amount of mineral supplement one should ingest is also not clear yet. A fair amount of scientific knowledge is, however, available about vitamins, so let us discuss these.

The different vitamins are usually required in small amounts only, and these amounts are usually age dependent, and have different bodily functions. Vitamin A aids in the conditioning of the skin and the mucous membranes. It also prevents night blindness and other diseases of the eye. A group of vitamins called B-complex vitamins prevents various diseases like beri-beri and pellagra. They also aid in body cell metabolism (the conversion of food into energy in body cells). Vitamin C prevents the disease scurvy. In recent years, it has become a center of controversy because of its possible role in preventing the common cold. Vitamin D is necessary for the growth of bones and the prevention of rickets. However, in excess, it is toxic. Vitamin E is associated with antisterility and has been shown to slow the aging process in laboratory animals. Vitamin K is essential for proper clotting of blood.

Recommended vitamin and mineral allowances by the National Academy of Sciences are given in Plate I.

It should be observed that mass ingestion of vitamins can be harmful. Large vitamin intake affects different people in different ways and the same person differently at different times. For example, large quantities of vitamin C can be tolerated readily by some individuals, while for others it may cause kidney problems.

For the latest recommended vitamin and mineral allowances, log on to http://vm.cfsan.fda.gov/dms/wh-nutri.html

Aging: Theory and Practice

Aging has been of universal interest throughout history. The cause of aging is not known. Isolated experiments relating to the process have, at times, yielded startling results. For example (*New York Times,* October 29, 1961, page E 7), two groups of chickens at the particular ages of 7 to 10 days were fed almost identical diets, with the exception that from one group was withheld a particular dietary substance called tryptophan. The group that was fed the incomplete diet ceased to age for periods of 6 to 9 months. When the tryptophan was restored, the chickens that had ceased to age then resumed the normal aging process.

In medical jargon, such isolated experiments are called *anecdotal* and are usually disregarded because of the lack of supporting statistical evidence. Nevertheless, anecdotal evidence of this type may indicate our ignorance of important underlying processes and may be worthy of deeper experimental study.

In recent years a variety of theories of aging have been proposed, some of which are as follows:

(1) Aging is encoded in each person's genes, that is, in those reproductive and hereditary cells that contain all the information about one's skin color, hair texture, muscle texture, bone size, and so forth. One of the findings that seems to support this theory is that mongoloids, who are known to have a specific genetic defect, age more quickly and die younger than people without this defect.

(2) Another theory is that aging is the result of the accumulation of waste products in cells. A living cell that cannot perform its functions simply dies. Indeed, it has been shown that protein cross linkages do form with age and that these hinder the efficiencies of cells. Also, a yellowish substance

called lipofuscin, which does not dissolve easily and appears to have no cellular role, does accumulate in cells with time.

(3) Other theories include such factors as diet, cell damage, disease, environmental factors, and error accumulation in cellular activities.

It may be that none of the above theories is correct. It also may be that some combination holds the answer.

In any event, even though the maximum age that can be attained has not yet changed, more people are now living longer, more active lives than ever before. This is a direct result of advances in science. The elimination of almost all contagious "childhood" diseases is a consequence of our knowledge about bacteria and viruses. Once debilitating or lethal diseases like tetanus, typhoid fever, syphilis, and rabies can now be cured by antibiotics or prevented by vaccines. Understanding of dietary needs has led to improved health and greater resistance to disease. Scientific study of exercise enables each person to improve and maintain well-being. In connection with exercise, a general maxim is "Use it or lose it." Eyes that are subjected to dark areas too long lose part of their ability to see. Arms and legs that are rarely used exhibit a decrease in muscle size (atrophy) and a loosening of the skin. A heart that is not exercised will fail more readily than one that is exercised. (See: *A Good Age* by Alex Comfort, Crown, New York, 1976.)

Note that as yet no antibiotic exists that cures viral diseases, although products exist that can shorten the length of a viral infection by preventing viral replication. This is the case, for example, with herpes. Currently available antibiotics are often prescribed to prevent bacterial infection, which can follow viral infection.

Conditions that at one time inevitably resulted in death

can, today, often be remedied by application of simple scientific principles. Consider, for example the case of a heart-attack victim who falls to the ground, stops breathing, and has no audible pulse. Without intervention, such a person will suffer irreparable brain damage within about five minutes and then will die. A simple method for keeping the person alive is called CPR (cardio-pulmonary resuscitation). It consists of rhythmically pressing down on the breast bone (sternum), which keeps the heart pumping blood, and blowing periodically into the person's mouth to provide oxygen. The method is taught at no cost at various Red Cross and Heart Association centers and takes about four hours to learn.

The Drive for Survival

The universal interest in keeping one's self and one's loved ones alive and healthy seems to be founded on a fundamental biological drive, which exists in all animals and in all plants: the drive to survive. Every action and, indeed, human history itself, can be interpreted in terms of this fundamental biological force. Survival is the reason why food, clothing, and shelter are basic human needs. It is the reason for sexual activity, which culminates in self-survival through the duplication of the parents' cells by reproduction. It is the reason for wars when a society cannot feed its people. It is reason for the religious beliefs of those who need assurance of survival after death.

The drive for survival manifests itself on two levels, in activities of self-survival and in activities of group survival. Of these, the more powerful is that of group survival, but usually the more active and readily observed one is that of self-survival. Self-survival is often viewed as self-concern or even selfishness. Self-survival is natural because each person views the world as if he or she were the center of everything, the reason being that our senses turn all outward perceptions inward.

In every society that has survived, laws have protected the needs of the group from misdirected actions of individuals. This is essential for the survival of the group, which can provide greater protection against destructive forces than can be provided by an individual acting alone. If each person does not follow the rules of society, crime and mayhem are natural consequences. Indeed, hunger can turn all but saints into uncivilized animals. However, it is only when the rules for, and actions of, the group assure individual survival that these rules are relatively safe from violation.

Often the drive for survival is either misunderstood or misinterpreted. For example, historians usually neglect it entirely and interpret history in terms of economic, social, or political forces, each of which is a manifestation of the underlying biological force. In daily life, the male drive to dominate a female, or many females, is rarely viewed as a natural force. Indeed, in all mammals, the male is driven for reproductive purposes by a herding instinct, through which he tries to keep females within his territory. (See: *The Illustrated Origin of the Species,* by Charles Darwin, abridged by R. E. Leakey, Hill and Wang, New York, 1979).

Finally, let us note that uncontrolled human reproduction is that specific manifestation of the survival drive that is the basis of many of today's world problems. Pollution, energy demands, housing shortages, harmful food preservatives, traffic congestion, smog, urban squalor, and food shortages are all, to a greater or lesser degree, the results of overpopulation. The means are available to control population growth. Their use is often contrary to the drive for individual survival. Nevertheless, were the existence of the entire human race to be threatened, it is possible that population growth would cease by both forced and natural constraints.

The Theory of Evolution

Lastly, let us discuss that biological theory about which almost all adults usually have a strong opinion, namely, the theory of evolution. However, before we begin, some preliminary remarks must be made. First, the theory does not maintain that humans are direct descendants of apes. Second, the theory of evolution is just that, namely, a theory. Finally, note that the theory is based on the scientific observation that all things change with time. Since this last observation is fundamental, let us expand upon it next.

The fact that we, ourselves, change with time is obvious to each of us. The reason is that changes in appearance, for example, are easily discernible by simply looking into a mirror. Many other changes in Nature are also easily observable, like how wind changes the shapes of clouds, how the availability of light changes the size of the pupils in our eyes, and how a forest fire changes the landscape. Other changes, however, occur slowly and may not even be observable during one's lifetime. These include the formation of canyons by flowing rivers, the movement of continents, and the orbital changes of planets. Our knowledge of such changes is usually dependent on the records of previous generations.

Let us return then to the theory of evolution. In terms of human history, the theory is relatively new, having been formulated independently by both Charles Darwin and Alfred Russell Wallace in the 1850s. Both Darwin and Wallace were naturalists. Darwin's name is usually associated with the theory because he wrote a book, *The Origin of Species*, on the subject, which was very popular and was read extensively.

It is perhaps easiest to approach the theory by observing that, paradoxically, all people are very much alike, yet very much different. We all have the same number of bones, all the bones are arranged in the same order, we all walk upright, we all have brains, hearts, and lungs. And yet, no two faces, no two voices, and no two fingerprints are exactly the same. There is a similarity in total design, but always a difference in details. This remarkable existence of both similarity and dissimilarity between two humans exist throughout Nature between any two animals and any two plants of the same species. Every two bluebirds and every two oak trees are both similar and dissimilar. Both Darwin and Wallace theorized primarily about the similarity of total design.

Darwin and Wallace began with the assumption that all things change with time. They then tried to explain how and why change occurs, and their answer was given as follows. They claimed that all animals and all plants have evolved from simpler forms of life by a process of natural selection, and that only the best adapted to the natural environment, survived and reproduced. Thus, Nature acted as a selective force in which only the healthiest survived disease, only the strongest survived famine, and only the fastest ourtran the predators. Thus, the development of all types of animals and plants was theorized as a dynamic process of improvement, which, by and large, was a slowly changing phenomenon.

The evidence in support of evolution is massive. Darwin and Wallace gave their own meticulously gathered evidence using birds and insects. Since then, evidence has been produced, for example, from fossil studies, genetic studies, and studies in comparative physiology. In this small space, we cannot begin to present all the positive arguments that support the theory. We will then concentrate on only a few arguments, which are considered to be fundamental, some of which also indicate gaps in the theory.

The first forms of evidence with regard to human evolution are in the bones of people who lived long before us. How long before us? No one really knows. Estimates of archaeologists put our predecessor's emergence at about 50 million years ago. As nearly as we can deduce, the bones indicate a decrease in skull size, an increase in brain capacity, and an increase toward vertical posture. (See: *The Course of Evolution,* by J. M. Weller, McGraw Hill, New York, 1969.)

We would like to have more evidence to fill the gaps. However, finding human remains from the distant past is extraordinarily difficult. Not only were there few humans in existence in the distant past, but the numerous changes in the earth's surface may have covered up possible evidence. Nevertheless, if evolution theory is correct in assuming that life evolved from simpler forms, then there should be ample evidence of the existence and evolution of simpler life forms. And this is correct. As the earth cooled and land masses formed, these land masses captured simple sea life and kept it stratified in the earth in a chronological order. These sea fossils are abundant and show clearly evolutionary changes over millions of years.

Next, one might ask a most fundamental question about evolution, that is: Where did living things come from? Again, we are not sure, since no one was around to observe what really happened. However, in recent years scientists have been able to produce in the laboratory the fundamental building blocks of living matter from nonliving matter. Experiments that have proved successful replicate an earthly primordial environment and from it produce basic life structures. Two of the most impressive results were obtained by Stanley Miller and Leslie Orgel. Miller considered an environment that was hot, humid, and gaseous. In it he allowed for the passage of electrical currents. Orgel considered an ice-age environment. Both experiments led to the creation of living-cell structural units from inanimate materials. Thus, life could have started at a time when the earth's atmosphere was hot and its surface was largely water, or it could have started during some ice age, or both. Again, we do not know. As scientists we only know that these experiments support the idea that life could have started in either way from inanimate materials.

Next, let us examine the wondrous process of human conception and birth. The human fetus develops from the fertilization of a single cell, the human egg. During its first eighteen days of development, it is similar to the embryo of any other animal that has a back bone (vertebrate). During its first four weeks, it has gill-like arches. During its first eight weeks, it has paddles and a tail. (See: *Behold Man,* by Lennart Nillson, Little, Brown and Co., Boston, 1973.) Only after two months is it recognizably human, even though the tail, which has decreased in size, persists. It lives in a warm watery environment for its first nine months. For its first year after birth, it crawls on all fours. And finally, when it is about one year of age, its spine attains the proper curvature so that it can and does stand erect and walk on two feet, like all other humans. This entire developmental process is remarkably similar to an evolutionary process.

Finally, we remark on one of the most significant discoveries of the last century. In the 1950s, Francis Crick and J. D. Watson deciphered the structure of DNA (deoxyribonucleic acid) in living cells. (This discovery and its significance will be studied in detail in the next chapter.) For the present let us only indicate that DNA is universal, occurring in all living cells, both plant and animal. Moreover, its structure varies only slightly from living cell to living cell. These slight changes over the centuries also support the contention that evolutionary changes are indicated by concomitant changes in a cell's DNA.

When one examines all the accumulated evidence in favor of the theory of evolution, the entire mosaic, though circumstantial and full of open questions, presents a compelling picture. The theory is difficult to refute. Indeed, it may be wrong. However, to date, no other explanation is as consistent with experiments and observation in explaining how life came to be as it is now.

III

Genetics—the Study of Population and Cell Heredity

The Old and the New Genetics

One of the branches of biology in which spectacular results have been achieved in recent years is genetics. Genetics began as a scientific discipline through the work of the monk Gregor Mendel, who experimented meticulously with pea plants. The hereditary principles that he discovered have been applied so successfully in crossbreeding that the production rates for beef, dairy products, and food crops now exceed all previous expectations.

However, in the 1950s, not only was a full understanding achieved of the inheritance materials, the genes, but some ability to manipulate them was also developed. The potential application of this new knowledge and these new techniques may also exceed all expectations.

Let us then examine both the older and the newer genetics in somewhat greater detail.

Population Genetics

Darwin and Wallace theorized about the similarity of total design in plants and animals of the same species. They did not consider why any two animals, for example, of the same species are somewhat different. If one thinks about humans, then physical differences that exist between any two people are readily understandable. First, observe that differences do exist between two children in the very same family. These differences need not be very great, but they do exist. Next, observe that particular family traits are rarely inbred, because most societies do not condone incest, thus encouraging diversity. Finally, add in those evolutionary changes that have resulted in different parts of the world due, for example, to different climates, and the individual differences that exist throughout the human race seem to be quite reasonable.

Now, children inherit physical traits from their parents. In fact, in all of Nature, whether it be with plants or animals, offspring inherit from parents. It is fundamental to understand how this inheritance works. However, if only a few traits were involved, it would be easier to grasp the fundamental principles. Historically, such a simple approach was taken in the middle of the nineteenth century by Gregor Mendel, who experimented with hundreds of garden pea plants and made astute observations about such simple traits as height and color. From these observations he discovered two fundamental laws of inheritance.

The first of Mendel's deductions was a basic one. It was that inherited characteristics are transmitted in individual, indivisible packets, now called genes. The experiments that led to this conclusion were very simple and can be characterized as follows. Assume that all pea plants under study fall into two types, those which are 2 feet tall and those which are 4 feet tall. Experiments reveal that a crossbred offspring, that is from parents of different heights, is never 3 feet tall. The plant is either 2 feet tall or 4 feet tall. This means that the contribution from one parent dominates that of the other parent. There is no averaging of parental contributions, from which Mendel's first principle was deduced.

Mendel's second discovery is the one used in agricultural crossbreeding. Within experimental error, he found that 3 out of every 4 offspring were always of the same height, and this height was the same in all experiments. In our example, say this height was 4 feet. Then the 4-foot height would be called a dominant trait and the 2-foot-height a recessive trait. It is perhaps curious that recessive traits appear at all. They do appear, however, only when a plant receives 2 recessive genes from its parents, which can be shown to be the case in 1 out of 4 chances.

Contemporary artificial breeding techniques, which rely heavily on this second discovery, have resulted in crossbred strains (hybrids) of animals and plants that incorporate, simultaneously, a number of superior qualities, which otherwise would not have resulted from natural breeding processes. (See: *Agricultural Genetics,* by J. L. Brewbaker, Prentice-Hall, Englewood Cliffs, 1964.)

Cell Genetics

Genetics is important in all forms of reproduction. Reproduction involving two parents is called sexual reproduction. That involving only one parent is called asexual reproduction. Though asexual reproduction is common in Nature, one of the most important places it occurs is in the cells of our bodies. Let us then examine the genetics of cellular reproduction.

All animals and plants are composed of cells. The simplest type of cell, which can be observed readily by placing a thin slice of onion under a microscope, has the shape of a rectangular box, which is, indeed, why it was first called a cell. Inside the cell is a liquid solution, which is about 85 percent water. Suspended in the liquid solution are many small particles and one relatively large, dense, oval-shaped particle called the *nucleus.*

In the human body, the variety of cells is more dramatic and diverse than in an onion. The shape of each human cell is adapted to a particular human function. Thus, a muscle cell is long and narrow and stretches easily, while a bone cell has many branches in it for structural stability. But, in every cell of any type, there is always a nucleus.

In the human body, it is essential that cells be able to reproduce. There are two reasons for this. First, the process of growth requires an increase in the number of cells. Second, many body cells cease to function at some time, either by nature or due to accident, and replacement is needed. Rubbed off or burned skin cells need to be regrown to protect the body from infection, whereas bone cells, for example, live only for about two years and need to be replaced if life is to be sustained.

Those body cells that can and do reproduce do so by a process called biological fission, in which a single cell divides into two cells, each of which is identical to the original. This fundamental process was studied extensively before 1950, but why a particular cell always reproduced as one of its own kind was never understood. Extensive research explored the question of why a blood cell reproduced as a blood cell, a bone cell as a bone cell, and a muscle cell as a muscle cell. The answer was provided in the 1950s and was of such generality that it applied to all animal and all plant cells. The answer lay in the nucleus. It was discovered by Francis Crick and J. D. Watson, who showed how genes, in the form of DNA, or, more fully, deoxyribonucleic acid, are transmitted and expressed in cell reproduction. Let us describe this result as simply as possible.

First, a few definitions are in order. A circular *cylinder* is a geometric shape like a smooth can (see Plate II (a)). A helix is a curve on a circular cylinder that winds up and around the cylinder at a uniform rate (see Plate II (b)).

What we describe next is valid for any type of cell, but in order not to be overly general, let us consider only a muscle cell. Our muscle cell consists of a *nucleus* and other cellular constituents (see Plate III (a)). Within the nucleus are long strands of material called chromosomes (see Plate III (b)). Within the chromosomes, among other things, are the *genes.* The genes encode the characteristics of our muscle cell through a double helix arrangement of the four substances *adenine, guanine, cytosine,* and *thymine,* that is, each gene consists of two helices in which each of the four given substances is bound to one of the other substances (see Plate IV). When the cell divides, the double helix separates and each part moves to a separate section of the cell. In addition, each helix replicates to become two double helices exactly like the one in the original undivided cell. The division into two cells (see Plate III (a)-(e)) is accompanied by the DNA in the divided parts transmitting messenger RNA with the genetic code to the rest of the cell, which then replicates as a muscle cell.

Current research is in the areas of transplantation and gene manipulation. The cure of genetic diseases is being explored actively, as is the possible role of damaged DNA in the development of cancer, which is a condition of uncontrolled cellular replication. However, almost fictional types of results can be produced by genetic manipulation. For example, insertion of a particularly firefly gene into a plant has been made to yield a plant that glows like a firefly. (See *Science News,* November 15, 1986.)

Other aspects of genetic research are being debated vigorously from ethical and moral points of view. In particular,

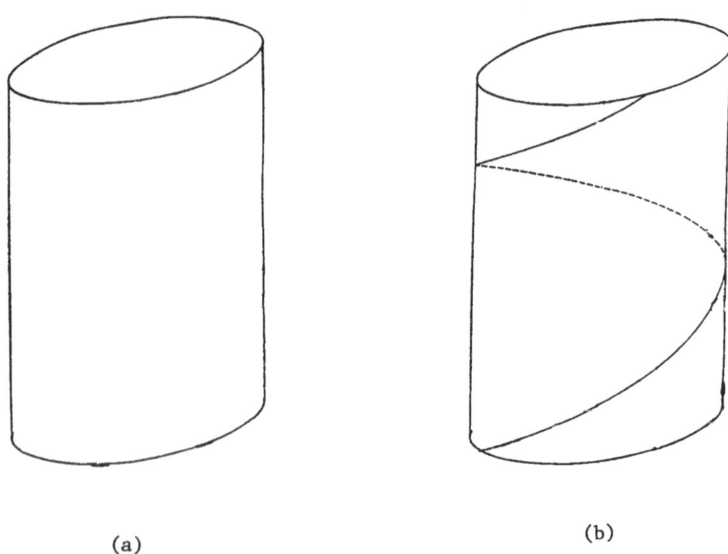

Plate II. **(a)** A circular cylinder.
(b) A helix on a circular cylinder.

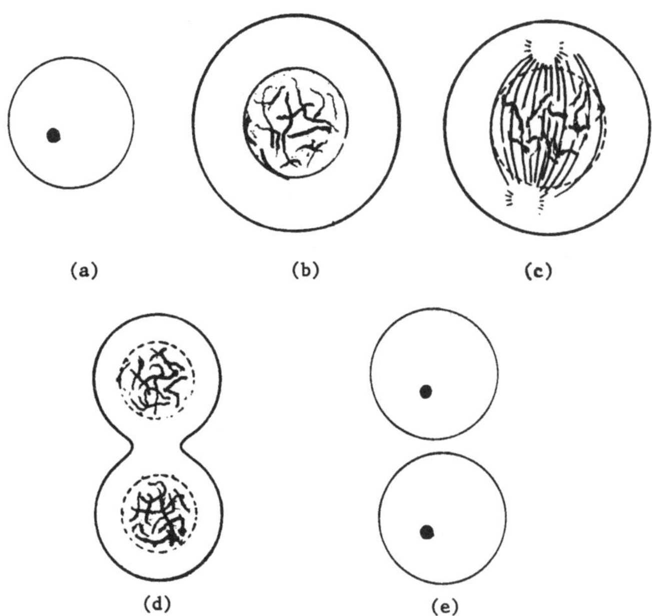

Plate III. The asexual reproduction of a cell, showing (a) the original cell, (b) the chromosomes, (c) alignment by the spindles, (d) near reproduction, (e) completed reproduction.

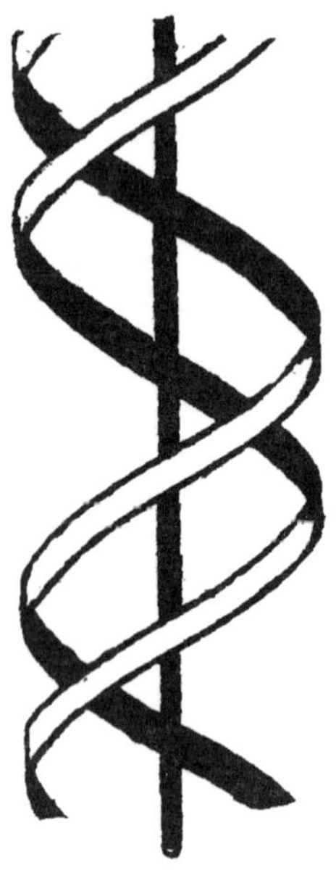

Plate IV. A section of a double helix.

cloning is a highly debated and debatable procedure. Cloning is the asexual reproduction of a plant or animal with no genetic changes, that is, it is an exact reproduction. The process usually involves the transfer of a nucleus from a body cell into an enucleated egg cell. In animals, some of these experiments have proved to be successful with frogs, rabbits, and sheep, the most famous of which is, perhaps, the sheep named Dolly. However, the entire process most often has not been achieved successfully and has yielded animals with various deformities. The question to be decided by society is whether or not *human* cloning is ethical or even desirable.

A more recent area of controversy concerns stem cell research. A stem cell is a cell in its most primitive state, that is, a cell which has not developed a specialized function. Nerve cells and blood cells, for example, are specialized cells which have developed from stem cells. Stem cell research of a medical nature aims at developing specialized cells from stem cells for cell therapy. Thus, for example, the development of nerve cells from stem cells would provide physicians with the means to replace nerve cells in individuals with nerve degenerative diseases, like Parkinson's disease and Alzheimer's disease. Stem cells are readily available in undeveloped human embryos, which has resulted in controversy over related research. Stem cells which are encoded to develop into blood cells exist in each individuals bone marrow.

IV

Chemistry—the Study of the Composition of Substances and the Transformations That They Undergo

The Scope of Chemistry

Chemistry is a subject that seems to deal with real magic. For example, if some vinegar is placed on litmus paper, which is blue, the paper turns red. There is no sleight of hand or any kind of chicanery involved. The paper's color is transformed from blue to red naturally. This transformation is only one of the many natural changes in substances, which chemists study by exploring the composition and interactions of various solids, liquids, and gases. Indeed, the chemical industry, in all its aspects, is the largest industry in the world.

A chemist who studies substances that are associated with life processes is called an organic chemist. Organic chemists study, among other things, bacteria, vaccines, antibiotics, vitamins, hormones, enzymes, and natural insecticides. A chemist who studies substances not usually associated with life processes is called an inorganic chemist. Inorganic chemists study, among other things, fertilizers, synthetic fabrics, explosives, petroleum purification, cosmetics, flame retardants, metallic alloys, paints, soaps, cleansers, and drugs.

Benefits and Liabilities

Unfortunately, the enormous benefits derived from chemical research are too often offset by concomitant liabilities. In particular let us consider three examples of contemporary importance. First, consider the development of modern chemical fertilizers and insecticides. Both enable an increase in food production in order to feed an expanding world population. But, insecticide drainage into rivers and lakes kills fish and other wildlife, while fertilizer drainage has led to such vast plant growth in lakes that many lakes are actually drying up.

As a second example, consider the extensive use of modern antibiotics. Their value is undeniable in the treatment of disease. Yet, their use discourages the development of natural antibodies and has resulted in the development of new, virulent bacterial strains, which are immune to a variety of antibiotics. Also, the lack of long-term experiments and studies may result in future problems that are more dangerous to the individual than the problem for which an antibiotic was used. This proved true for users of sulfa drugs for minor infections in the 1950s, in which the long-term effect in many cases was the development of kidney stones.

As a third example, let us consider a major social effect of the use of antibiotics and birth-control pills. Antibiotics can be used to treat both syphilis and gonorrhea, two major venereal diseases. Birth control pills can be used successfully to prevent conception (though their long-term effects on the circulatory and reproductive systems are still unknown.) Elimination of the fears of both pregnancy and venereal disease resulted in a reversal of sexual mores, called the sexual revolution. The effect on society has been unbalancing. Individuals raised before 1950 were raised to respond emotionally to conditions that had been reversed by 1970, and many could not adjust to the new social structure. In addition, the new sexual freedom has led to the spread of new viral venereal diseases, namely, herpes and AIDS, for which no cures are available.

Atoms and Molecules

The study of chemistry begins with the elements. All known substances are made from about a hundred basic elements in Nature (see Plate V.) These include oxygen, carbon, hydrogen, lithium, uranium, and zirconium. In proper combination, the elements are the known constituents of all other substances.

The smallest unit of any particular element is called an atom. Like a biological cell, every atom has a dense, central particle called a nucleus. Unlike a biological cell, an atom has no liquid environment and no boundary wall. In addition, the nuclear constituents are quite different. Outside of every atomic nucleus are small particles of negative electricity called electrons, while inside the nucleus are, usually, a balancing set of positive electrical particles called protons. The difference between atoms of two elements is characterized completely by the number of electrons outside the nucleus. Thus, in the first 100 basic elements, the first has only one electron outside the nucleus, the second has two electrons, the third has three, and so on. The element numbered 100 has exactly one hundred electrons outside the nucleus. The element with only one electron is called hydrogen and is denoted by H. It is the simplest of all the elements. The element with exactly eight electrons is called oxygen, and it is denoted by O.

Now, water, which is the basic fluid of life, is not an element. The smallest particle of water consists of one oxygen atom and two hydrogen atoms, all neatly bound together by what is called a chemical bond. Water is aptly designated by the symbol H_2O, which indicates which atoms and how many of each make up a particle of water. The smallest possible particle of water is called a molecule. It, like other molecules, is simply a combination of two or more bonded atoms.

Suppose now that we decide to make some water. We might first look around in the air and capture an oxygen atom and two hydrogen atoms. We need next to entice the three atoms to form a molecule. If left alone, the atoms will not come together. One must actually exert some force, or some energy, to make them bond, and this energy goes into the molecule when they do bond. Quite naturally, it is called the bonding energy of the molecule. All molecules are held together by bonding energy. When a molecule is unbonded, it gives up this stored energy. When this is understood, one can begin to understand how chemical reactions in the body that yield unbonding enable us to derive energy from food molecules, and this will be explored next.

Chemical Production of Energy in Living Cells

From the mouth to the rectum is one long passageway, which many scientists consider to be part of the outside of the body. When a child swallows a coin or an individual eats food that is not easily digested, like uncooked kernels of corn, both the coin and various corn kernels traverse the en-

element	symbol	atomic number
Actinium	Ac	89
Aluminum	Al	13
Americium	Am	95
Antimony	Sb	51
Argon	Ar	18
Arsenic	As	33
Astatine	At	85
Barium	Ba	56
Berkelium	Bk	97
Beryllium	Be	4
Bismuth	Bi	83
Boron	B	5
Bromine	Br	35
Cadmium	Cd	48
Calcium	Ca	20
Californium	Cf	98
Carbon	C	6
Cerium	Ce	58
Cesium	Cs	55
Chlorine	Cl	17
Chromium	Cr	24
Cobalt	Co	27
Copper	Cu	29
Curium	Cm	96
Dysprosium	Dy	66
Einsteinium	Es	99
Erbium	Er	68
Europium	Eu	63
Fermium	Fm	100
Fluorine	F	9
Francium	Fr	87
Gadolinium	Gd	64
Gallium	Ga	31
Germanium	Ge	32
Gold	Au	79
Hafnium	Hf	72
Helium	He	2
Holmium	Ho	67
Hydrogen	H	1
Indium	In	49
Iodine	I	53
Iridium	Ir	77
Iron	Fe	26
Krypton	Kr	36
Lanthanum	La	57
Lawrencium	Lr	103
Lead	Pb	82
Lithium	Li	3
Lutetium	Lu	71
Magnesium	Mg	12
Manganese	Mn	25
Mendelevium	Md	101

element	symbol	atomic number
Mercury	Hg	80
Molybdenum	Mo	42
Neodymium	Nd	60
Neon	Ne	10
Neptunium	Np	93
Nickel	Ni	28
Niobium	Nb	41
Nitrogen	N	7
Nobelium	No	102
Osmium	Os	76
Oxygen	O	8
Palladium	Pd	46
Phosphorus	P	15
Platinum	Pt	78
Plutonium	Pu	94
Polonium	Po	84
Potassium	K	19
Praseodymium	Pr	59
Promethium	Pm	61
Protactinium	Pa	91
Radium	Ra	88
Radon	Rn	86
Rhenium	Re	75
Rhodium	Rh	45
Rubidium	Rb	37
Ruthenium	Ru	44
Samarium	Sm	62
Scandium	Sc	21
Selenium	Se	34
Silicon	Si	14
Silver	Ag	47
Sodium	Na	11
Strontium	Sr	38
Sulfur	S	16
Tantalum	Ta	73
Technetium	Tc	43
Tellurium	Te	52
Terbium	Tb	65
Thallium	Tl	81
Thorium	Th	90
Thulium	Tm	69
Tin	Sn	50
Titanium	Ti	22
Tungsten	W	74
Uranium	U	92
Vanadium	V	23
Xenon	Xe	54
Ytterbium	Yb	70
Yttrium	Y	39
Zinc	Zn	30
Zirconium	Zr	40

Plate V. Chemical elements (1975).

tire passage from the mouth to the rectum and emerge in almost the same condition as when they entered. Food that supplies energy to body cells, however, follows a different route.

A bulk of food usually enters the mouth (food in special forms can be fed directly into the bloodstream). The front teeth, called the incisors, merely cut smaller chunks from larger ones. The back teeth, or molars, break down the bulk further with the help of acid from saliva. The tongue then tosses the resulting ground-up mash into the stomach where it is further dissolved by hydrochloric acid. Stomach contractions then force the finely ground-up mash into the small intestines. In the small intestines are millions of porous, hairlike formations, which allow digested food particles to enter the bloodstream. Also, food particles still not sufficiently dissolved are acted upon further by enzymes, which speed up the process of chemical breakdown.

Once in the bloodstream, the ultra-fine food particles are transported throughout the body to be used by the cells. When they reach, say, a muscle cell, by some unknown means, thought to be electrical, the cell allows certain particles in and keeps others out. Once inside the cell, the molecules are unbonded with the aid of oxygen, thus releasing energy for the cell to do its work. After the energy has been used, the remaining chemical constituents are sent back into the bloodstream, where they are carried to the large intestine or to the kidneys. The kidneys are filters and allow the emergence of liquid waste from the body through the bladder. The large intestine and the bowels generally have bacteria that help solidify emerging waste products into stools, which then emerge through the rectum.

The knowledge that we have concerning the body's use of food has many gaps in it, even though an immense amount of knowledge is available. Once again, we do seem to have an understanding of the larger framework, but we do not know many of the exact details. We do know, however, that a proper chemical balance within the body is necessary not only for physical health, but also for mental health, and that proper diet and exercise help in the attainment of such a balance.

V

Physics—the Study of Matter, Energy, and Motion

The Three Branches of Physics

Physicists, chemists, and biologists have overlapping interests, especially in the study of atoms and molecules. But physicists are concerned with a number of other studies which are of a lesser interest to chemists and biologists, and include gravitation, light, heat, sound, electricity, magnetism, X rays, radio waves, lasers, black holes, chain reactions and quarks. Today there are three major areas of physics in which both theoreticians and experimentalists are active. These are Newtonian physics, relativistic physics, and quantum physics. Historically, they developed in the given order and each has a unique applicability, which we will discuss next.

Newtonian Physics

Early science was often intimately intertwined with religion and mysticism, from which scientists began a torturous path after the fourteenth century. In this breaking away, experimental scientists were aided by the availability of new, relatively powerful devices for observation, like the telescope. Theoretical scientists were aided by the development of a new, powerful mathematical discipline called the calculus.

One of the developers of the calculus was Isaac Newton, one of the most outstanding scientists of all time. Newton did foundational work in the studies of gravitation, light, and mechanics. He developed the first general equation to describe how things change with time. The equation was so general that it applied equally well to the motion of a falling apple and to the motions of the planets around the sun. The equation was called Newton's dynamical equation and was the cornerstone of physics until 1900.

Relativity

Toward the end of the nineteenth century, new phenomena were discovered in the form of radiated waves, like X rays, radio waves, and light waves. Further study revealed that these waves all had common features. For example, they all traveled at the incredible speed of 186,000 miles per second. These newly discovered waves were called electromagnetic waves and were studied collectively in physics in an area called electromagnetics. However, it was soon discovered that Newton's dynamical equation was incorrect in describing the motion of electromagnetic waves.

An entirely new mathematical formulation and a new dynamical equation were required. The work was begun by J. C. Maxwell and H. A. Lorentz and was completed by Albert Einstein in his special theory of relativity. Thus, the very first paper in relativity was devoted strictly to the study of electromagnetic waves. But, so powerful was this new theory that subsequent papers demonstrated its applicability to the study of distant galaxies, particles moving at high speeds, and atomic energy. Let us elaborate on this last application.

From the special theory of relativity follows an equation that is not a consequence of Newtonian physics. The equation is the now famous

$$E=mc^2$$

or, in greater detail,

$$E=m \times c \times c$$

or, again, as

E=m (*multiplied by*) c (*multiplied by*) c.

The equation means that any object that is at rest and that has a mass of m units has an internal energy E given by the above formula, where c is that incredible speed of 186,000 miles per second. As an example, an object on earth weighing *1* unit would have the immense internal energy of

$$E=1 \times 186{,}000 \times 186{,}000 = 39{,}204{,}000{,}000 \text{ units.}$$

If, in some fashion (to be discussed later) an object of 1 mass unit were to be annihilated, the above energy would be released.

It should be noted, however, that though relativity theory has very important applications, it also has severe restrictions. It could not and did not replace Newtonian physics entirely. For example, relativity theory cannot be applied to the motion of more than one body in the solar system. Thus, relativity applies to the motion of the planet Mercury around the Sun, but does not apply if the motions of both Mercury and Venus are to be analyzed simultaneously. Indeed, rocket trajectories and planetary trajectories must be and still are calculated using Newtonian physics.

Quantum Mechanics

In the early 1920s, still a third branch of physics was developed. Neither Newtonian physics nor relativity seemed to be correct when analyzing fundamental motions *within* atoms and molecules. The new subject that emerged was called quantum physics or quantum mechanics, and it was directed primarily to the internal workings of atoms and molecules.

Quantum mechanics is difficult to understand because, for most people, it is so contrary to their daily experiences that their intuition fails them. However, just as soon as one realizes that the radius of the smallest atom, that is, hydrogen, is only 0.0000000037 centimeters, then it becomes reasonable to see that the understanding of such a minute entity may not correspond to what our intuition expects.

The theory is founded on an assumption of discreteness. For example, in it one assumes that energy exists in discrete, individual packets called quanta. An atom can have 1 unit of energy, 2 units of energy, 3 units of energy, and so on, but it cannot have 2 1/2 units of energy. In a supermarket, one can by 2 1/2 pounds of potatoes, but in quantum mechanics, fractional portions of energy are not allowed. Next, one assumes that one does not have *exact* knowledge about the internal workings of atoms and molecules, but only *probable* knowledge. Thereby, the subject is dependent on the theory of probability. Thus one cannot know where an electron is within an atom and what its velocity is, simultaneously. One only knows the *probabilities* associated with position and velocity. In everyday life, if a car is in motion, we could easily determine both where it is and how fast it is moving. Quantum mechanics does not allow such a conclusion if one is describing the motion of an electron within an atom or molecule, but only allows us to determine the probability that an electron is at a given position. Thus the position of an electron in an atom has an associated *fuzziness.*

The full development of quantum mechanics enables one to calculate almost exact energies of many atoms and molecules and to explain why each chemical element, when heated, gives off a light with its own, very individual characteristics. In addition, many other results are most notable. However, quantum mechanics also has its limitations. For example, it has failed consistently to yield correct results for the energy of nuclei, and, like relativity, does not seem to be applicable for simulation of the motion of the solar system.

Nuclear Energy and Reactors

One of the most important areas of physical research today is that of energy. Let us then explore the energy source that is most studied by physicists and also most controversial in the public domain, nuclear energy.

One method of deriving nuclear energy is by splitting an atom's nucleus. When a nucleus is split, it gives up the vast amount of energy that held it together. The process is called nuclear fission. The aim of physicists for many years was to devise a method by which splitting a few nuclei, at the same time, would be self-perpetuating in that these split nuclei would, themselves, go on to split other nuclei. Such a reaction is called a chain reaction and would produce energy continually. It was first achieved in 1942, during World War II, by Enrico Fermi at the University of Chicago. The technological device in which a chain reaction is accomplished is called a nuclear reactor.

A chain reaction is usually achieved as follows. A very small particle, smaller than an atom, called a neutron, is fired at a large mass of uranium. Uranium is used as a target because it is readily fissionable material. The neutron is fired slowly in order to improve its chances of splitting one of the nuclei. If it does achieve its goal, then, fortunately, the split uranium nucleus not only gives up its energy, but it emits new neutrons, which go on to split other nuclei, and so on, forming a chain reaction.

Now, uranium exists in several forms. The kind used for the chain reaction described above is called U-235. The reaction with a slowly fired neutron is called a thermal reaction because of the vast amount of heat energy it generates. The reactor is called a thermal reactor. It is common to generate electricity using a thermal reactor in the following way. The heat energy is used to boil water. The resulting steam turns a turbine, which, in turn, drives an electric generator. Using only one pound of uranium, one can generate the electricity usually generated using one million pounds of coal. A thermal reactor used in this way also requires large additional amounts of water for cooling purposes.

If one uses another form of uranium, namely, U-238, then the slow neutron technique simply does not work well in achieving a chain reaction. However, neutrons fired at high speeds do work well. They do not set off as many chain reactions as in the case of slow neutrons fired at U-235, but they have a most interesting and important value. The fast neutron fission, when it does result, is accompanied by a chemical reaction in which the U-238 is transformed into plutonium-239. This plutonium-239 is itself, fissionable so that the process has given birth to a new source of fissionable material. A reactor that accomplishes this is called a breeder reactor.

Problems with Nuclear Fission

The basic problem with all types of nuclear fission processes is that they yield radioactive waves and wastes. Radioactive waves are like X rays. They pass through the body without our knowledge. However, like X rays, a large dose of such waves, as it passes through body cells, can disturb the cell contents so severely that the cells can die or reproduce unnaturally. Excessive radioactive damage of blood cells, for example, can lead to a form of blood cancer called leukemia. Too much exposure forms a health hazard, and, as yet, there is no agreement as to what levels of exposure are reasonably safe.

Unfortunately, the substances that are the waste products of fission are, themselves, radioactive, and they may not lose their radioactivity for thousands of years. We do not know at present how to deactivate these materials, so there is an effort to store them into the earth, where they may not be hazardous to plant and animal life. Efforts to deposit them in metal canisters at the bottom of the sea resulted in the discovery of leakages, so that plan was abandoned. If too much radioactive waste were to accumulate in a world full of reactors, the problem could result in the end of all life as we know it.

A second problem resulting from the fission process comes from the disposal of coolant water into rivers and lakes. This hot water has affected the natural ecology in every area where a thermal reactor has been built. Initially, it results in the death, for example, of fish that cannot adjust to the extreme changes in the water temperature that result from disposal. The long-range effects are, as yet, not clear.

Nuclear Fusion

A much studied "clean" alternative to nuclear fission is nuclear fusion, for which the practical technology has not as yet been developed. Nuclear fusion begins with the smallest atoms, that is, with hydrogen and helium, rather than with the largest ones, like uranium and plutonium. Consider, for example, two hydrogen atoms. Instead of just trying to bond the two atoms into a molecule, which is difficult enough, suppose we try the more difficult process of pushing their nuclei together into a single nucleus. This process is called fusion. It does not result in a molecule, but results in a larger atom.

If we can find the energy to push their nuclei together, then a very interesting phenomenon occurs. The mass of the resulting atom is smaller than the sum of the two individual masses with which one began. Thus, in combining two atomic nuclei into a single nucleus, some of the mass is left out. In fact, this mass is annihilated by conversion into energy, called the energy of fusion, and there are no waste products. The amount of energy so created is immense and is given by the Einstein equation $E=mc^2$.

It is thought that the Sun generates its energy by a fusion process.

VI

Astronomy—the Physics of Celestial Bodies

The Universe

One special field of physics that has a history of its own is astronomy. The wonders of the sky, the Sun, the planets, the stars, the moon, comets, meteors, and eclipses have captured human imagination through the ages. Knowledge of astronomy is also practical because it yields information, for example, about the time for planting, the time of flooding, and one's navigational course at sea.

Let us begin by exploring what is known about the size and nature of the cosmos, which is just another name for the universe. The Earth is one of nine known planets that revolve around the Sun. These planets, in order of their proximity to the Sun, are Mercury, Venus, Earth, Mars, Jupiter, Saturn, Uranus, Neptune, and Pluto. The Earth is approximately 93 million miles from the Sun. To understand the Earth's size and position relative to the Sun, one can think as follows. If the Sun were the size of a soccer ball, then the Earth would be the size of a golf ball and the golf ball would be about 600 feet, or 183 meters, away from the soccer ball.

Our Sun, like all suns, is a star. It is, however, a relatively small star. The sun that is nearest to ours is named Proxima Centurai, and its distance away is 260,000 times the distance between Earth and our Sun, or, approximately, 24 trillion miles away. If it were possible to drive, at 55 miles per hour, to Proxima Centurai, the trip would take more than half a billion years.

Our Sun and Proxima Centurai are only two suns in a collection of over 100 billion suns, which are all rotating as a system called the Milky Way galaxy. When one looks up into the night sky and identifies the myriad stars of the Milky Way, one is looking into the interior of the galaxy from a position near the outside. The size of our galaxy exceeds all usual intuition and comprehension.

But, further, our galaxy is only one of at least one-hundred million other galaxies. None of these can be seen by the naked eye alone. Some can be seen with a small telescope. Many others can be observed readily with large telescopes of both the optical and radio types. Our galaxy, as large as it is, occupies no more than one trillionth of the universe described thus far.

Whether even more exists is open to speculation. But what has been described already is sufficient to establish that man is a mere speck in the cosmos and a life span of a hundred years is a mere instant in cosmological time. The recent space probes into our solar system reach out almost no distance at all into the space of the universe. These observations often help to cultivate personal humility.

Mankind's Place in the Universe

It has taken humans over two-thousand years to accept the fact that we are not the center of all that exists. This change of attitude resulted after much bitter controversy between science and religion. The problem began with the crude astronomy developed by Ptolemy during the Golden

Age of ancient Greek civilization. Though this primitive form of astronomy assumed that all celestial bodies rotated around Earth, one could actually use it to make some relatively simple, almost accurate predictions about astronomical phenomena. Subsequent Western culture inherited and accepted this theory, which was consistent with the dominant religious beliefs of the day. After all, since humans were deemed to be the supreme life form created by God, it was appropriate that humans should be located at the center of God's universe.

In the early 1500s, the Polish astronomer Nicolaus Copernicus showed that Ptolemaic theory could be improved immeasurably by placing the Sun at the center of the solar system. This new theory removed Earth, and life on Earth, from the focal to a secondary position, which was contrary to religious doctrine. However, Copernicus never witnessed the response of the church, for he was seventy years old when he submitted the theory for publication and he died before it appeared in print.

After the death of Copernicus, two other eminent scientists, Johannes Kepler and Galileo Galilei, took up his work. Kepler showed that, in fact, the motions of the planets were not circular, but more oval, or, to be precise, elliptic. Kepler derived three fundamental laws of planetary motion within a Copernican system. These have stood the test of time through today. Interestingly enough, these laws were submerged in a theory about the music of the celestial spheres and about angels in heaven. One might suspect that this was Kepler's way of disguising results that he knew would be socially, politically, and religiously unacceptable. However, a reading of the manuscript gives the impression that Kepler believed all he had written.

Galileo Galilei was, perhaps, the most eminent scientist in Italy from the late 1500s to the middle of the 1600s. His use of experiment and theory to complement each other marked the beginning of the modern scientific method. He studied a variety of types of physical phenomena. With regard to astronomy, he developed experiments with gyroscopes to simulate planetary rotation. But it was his improvement and use of the telescope that provided him with the final evidence to believe Copernican theory. For, through his telescope, he saw the moons of Jupiter and observed that these moons rotated around Jupiter, not around Earth. Thus, the basic assumption of Ptolemaic theory, that all celestial bodies rotated around Earth, was now known to be false. To the naive, it does not seem possible that the political and church authorities could prevail over Galileo's ensuing and assured support of Copernican theory. But the response came quickly. Galileo was subjected to an inquisition as a heretic and was threatened with death by torture. At an age close to seventy, this great scientist recanted all support of Copernican theory and was sentenced to house arrest for the remainder of his life, which lasted nine more years.

The trial of Galileo had a devastating effect in all European countries dominated by the Catholic Church. Science in all the southern countries of Europe went into decline. Noted scientists even emigrated. But, the flame was taken up in the northern countries, such as England, where Isaac Newton was soon to begin his researches. The later triumph of Newtonian physics supported the validity of Copernican theory and assured its final acceptance.

Cosmology in our Daily Lives and Thinking

Finally, let us make a few remarks about the effects that the universe has on our daily lives. There is no doubt such effects do exist. For example, solar storms, solar prominences, and cosmic rays upset the electrical fields of our atmosphere, which does affect electrical reception and other aspects of our lives (see Plate VI). We are not certain to what extent such phenomena affect us, and research in this area is in progress. However, there is absolutely no evidence that these phenomena are related to when or where a person was born. Thus, there is absolutely no evidence to support any aspect of astrology.

There is also no scientific evidence to support UFO (Unidentified Flying Object) activities from outer space. This does not mean that all people who report having seen a UFO are either irrational or are making fraudulent claims. For example, aberrations of light can distort visual perception. Also, an uninformed public will frequently report an experimental aircraft, like a flying platform, to be a UFO.

There is also a high probability that life exists elsewhere in the universe, but it is equally probable that it is of a form completely different from life as we know it.

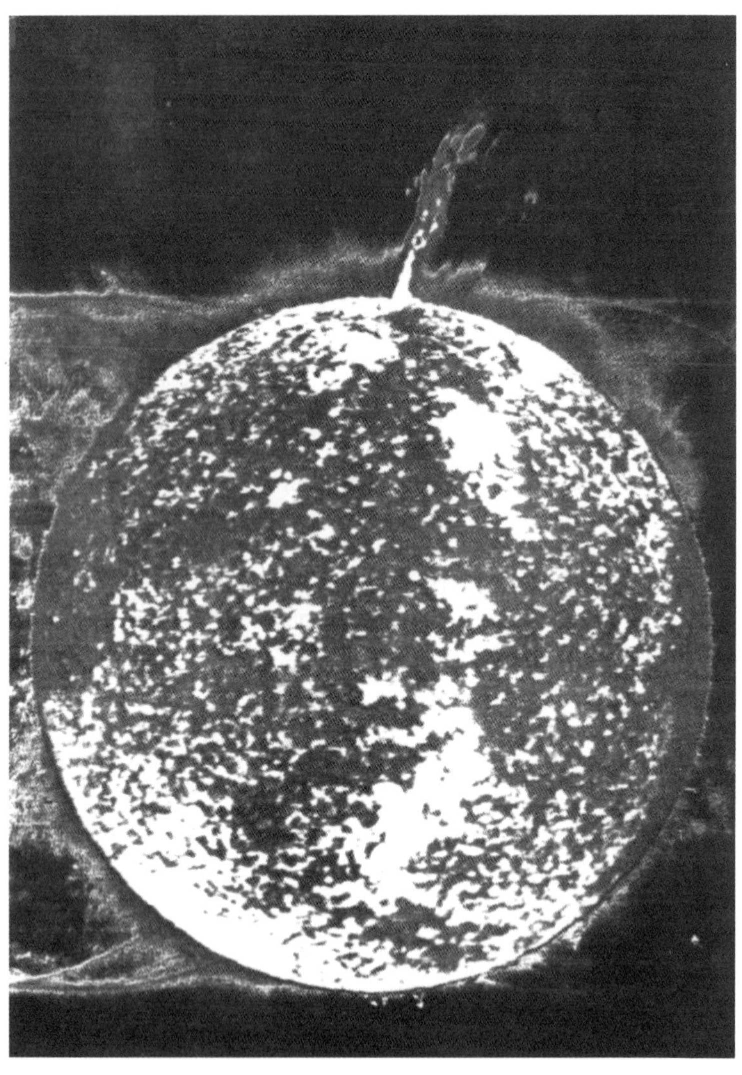

Plate VI. A view of the sun from Skylab, showing a prominence that has erupted into space.

Epilogue

An understanding of Nature is essential if we are to make wise choices about the future of the world in which we live (see Plate VII). No one science stands alone in helping us to understand Nature. For example, biology helps us to understand how plants make food molecules, physics enables us to calculate the energy in such molecules, and chemistry shows us how that energy is released for use in the body. There is a unity of science that is essential for complete understanding. Without this unity, our understanding can never be complete.

The fear of scientific illiteracy expressed in 1980 by the *New York Times,* and quoted in the Prologue, should be of even more concern today. There are two primary reasons for this. First, modern computers were not introduced into scientific work until the early 1950s and have since resulted in an explosion of knowledge in all the sciences. Second, cell genetics did not begin its development until the middle of the 1950s, and it is still in a period of unparalleled growth.

Finally, we must observe that the achievements of science are all consequences of the vast capabilities of the human mind, and that these capabilities are the natural gifts of every man and woman.

Plate VII - Cartoon by Golliver.

Part Two

A Dictionary of Scientific Terms and Words

A

aborigines, n. Usually, the native people of Australia, who preceded settlement by the English.

abortion, n. The expulsion or extraction of a mammalian fetus before it is viable.

abscess, n. A sore on a membrane tissue that discharges pus and originates deeply within the tissue.

AC. (See: alternating current).

accelerator, n. In physics, any device used to increase the speed of a charged particle, like an electron; in chemistry, any substance that increases the speed of a reaction.

acetate, n. A salt of acetic acid.

acetic acid. A common acid compound that forms about 10 percent of vinegar.

acid, n. Any substance that dissolves in water, has a sour taste, and turns litmus paper red; any compound that dissociates in water with the production of hydrogen ions.

acid rain. Rain with a relatively high content of sulfuric acid and nitric acids, which have formed from the chemical interaction of water and pollutants in the air.

acrophobia, n. Fear of being in relatively high places.

actin, n. A basic muscle protein needed for contraction and relaxation.

acupuncture, n. A technique developed in ancient China for relief of pain by the insertion of needles at key points of the body.

adenoids, n. An abnormally enlarged mass of lymphoid tissue, which forms an obstruction in nasal passages.

adenosine triphosphate (ATP). A chemical of importance in the conversion of food into energy in the body cells.

adipose tissue. Tissue with stored fat.

adrenaline, n. An adrenal gland secretion that increases the heart's contraction rate, dilates bronchioles, and increases blood sugars, thereby allowing vigorous physical action.

aerobic, adj. In biology, living or active only in air.

aerosol, n. A fine liquid or solid suspension in air or gas.

aether, n. (See: ether.)

Agent Orange. A toxic herbicide.

ague,n. A condition marked by fever and chills.

AIDS (acquired immune deficiency syndrome). A disorder of the body's immune system caused by infection with HIV (human immunodeficiency virus).

airfoil, n. Any winglike surface designed to obtain a lift as it moves through air.

air pollution. The contamination of pure air by substances that make breathing difficult or hazardous.

albinism, n. A deficiency of pigment in the skin, hair, and eyes.

alchemy, n. A medieval science devoted to transforming base metals into gold and to developing the means of prolonging life forever.

algae, n. A class of primitive, plantlike organisms often found in water.

alimentary canal. The tubular passage of the body from the mouth to the rectum.

alkali, n. A type of mineral salt.

alkaloid, n. A class of nitrogen-containing substances, including morphine, nicotine, and quinine, noted for their powerful physiological effects.

allergy, n. Any excessive physical sensitivity to certain substances like pollens and medications of various types.

allotropy, n. The capability of certain elements to appear in diverse forms, like carbon, which can take the form of a diamond or a charcoal.

alloy, n. Any substance composed of two or more metals.

alpha particle. A nucleus of a helium atom moving at a very high speed.

alternating current (AC). A type of electrical current characterized by oscillating electron motion, rather than motion in a fixed direction.

alveoli, n. The small air spaces in the lungs where carbon dioxide can exit into the blood and oxygen can enter from the blood.

Alzheimer's disease. An incurable, progressive, degenerative brain disorder of unknown cause that produces gradual loss of memory and other intellectual functions.

amino acid. A fundamental group of more than twenty acids that are the chief components of all proteins.

ammonia, n. A pungent, volatile gas composed of hydrogen and nitrogen.

amoeba, n. A microscopic, one-celled animal that lives in water.

amphibian, n. A vertebrate class of animals between fish and reptiles, including frogs, toads, and salamanders, characterized by an early aquatic stage, which has gills and an adult stage, which has lungs.

analgesic, n. Any drug that relieves pain.

anemia, n. A physical condition marked by a deficiency of red blood cells or hemoglobin, and often accompanied by paleness, shortness of breath, and heart palpitation.

anemometer, n. An instrument for measuring wind velocity or force.

anesthetic, n. Any drug or gas that induces partial or total loss of feeling or sensation.

aneurysm, n. An abnormal, blood-filled dilation of an artery, which results from disease of the artery wall.

angina pectoris. Muscular spasms of the chest, which are related to the functioning of the heart.

anodyne, n. Any drug that relieves pain.

anoxic, adj. Pertaining to a deficiency of oxygen.

anthrax, n. A highly infectious, usually fatal, bacterial disease of cattle and sheep, occasionally found in other animals, including man.

anthropoid, n. A class of animals closely resembling man.

anthropology, n. The study of the physical and environmental development of man.

antibiotic, n. Any chemical substance produced synthetically or naturally by a microorganism, especially a bacterium or a fungus, which kills or debilitates other microorganisms.

antibody, n. Any natural substance produced by the body that prevents infection or disease.

anticoagulant, n. Any substance that prevents clotting of the blood.

antigen, n. Any substance that, when introduced into the body, stimulates the production of antibodies.

antihistamine, n. Any substance that prevents or retards the secretion of histamine.

antimatter, n. A substance of the exact opposite nature of matter, which, when combined with matter, results in total annihilation and conversion into energy.

aorta, n. The main artery that leads away from the heart.

aphasia, n. Loss or impairment of the facility to use or understand speech.

aphid, n. A type of plant louse.

aphrodisiac, n. Any substance that, when ingested, excites sexually.

apod, adj. Footless, or, as in the case of eels, without fins.

apsides, n. In astronomy, the points of a planet's orbit that are nearest and farthest from the sun.

aquiform, adj. Having the form of water; like a liquid.

arachnid, n. Any eight-legged, insectlike animal whose body consists of two sections, like a spider, scorpion, or tick.

archaeology, n. The exploration and study of material remains of past human life.

arteriosclerosis, n. Any of several diseases of the arteries, all usually involving the formation of fatty or calcium deposits.

artery, n. Any of the large, tubular blood vessels that carry blood from the heart to the tissues.

arthritis, n. A particular type of inflammation of the joints, which is usually painful and often debilitating.

asbestos, n. A variety of silicon that occurs as fibers and is noncombustible, nonconducting, and chemically resistant.

asphyxiation, n. Death caused by deprivation of oxygen or air.

asteroid, n. A relatively small, planetlike celestial body in rotation around the Sun.

astigmatism, n. Inability to focus an eye due to an imperfection in the shape of an eyeball.

astringent, n. A substance that causes shrinking, contraction, or binding of tissues.

atherosclerosis, n. A form of arteriosclerosis in which fatty tissues form on the linings of the arteries.

atom, n. The smallest quantity of an element that retains the characteristic properties of the element.

atomic energy. (See: nuclear energy.)

atomic number. In chemistry, the number assigned to a given element; the number of electrons in one atom of a given element.

atrophy, n. A wasting away of part of the body due to lack of nourishment or use.

auditory nerve. The nerve that carries to the brain the impulse in the ear that causes hearing.

aurora borealis. Electrically produced streamers of light that appear in the northern skies from time to time; the Northern Lights.

autism, n. A psychological disorder involving absorption in fantasy and extreme activity and self-preoccupation.

avian, adj. Pertaining to birds.

axon, n. A fibrous extension that transmits impulses away from nerve receptor cells.

azimuth, n. A special section of the horizon relative to the observer.

B

bacteria, n. A large class of one-celled microorganisms of three basic shapes: spherical (coccus), rodlike (bacillus), and threadlike (spirillum), some of which are involved in beneficial chemical transformation processes, others of which are harmful in that they cause diseases.

bacteriology, n. The study of bacteria.

bacteriophage, n. A virus that infects and destroys bacteria.

barbiturate, n. One of a large group of drugs used as sedatives.

barometer, n. An instrument for measuring atmospheric pressure.

baryon, n. Any of several relatively heavy subatomic particles, including protons and neutrons.

basal metabolism rate, n. An index of the rate at which one's body uses oxygen in the conversion of food into energy.

base, n. Any chemical substance capable of reacting with an acid to form a salt; a compound that yields positive hydroxide ions in water; ammonia is a typical base.

bathyscaphe, n. A small, navigable diving vessel used for ocean exploration.

beri-beri, n. A disease characterized by general debility and painful stiffness, caused by insufficient vitamin B_1 in the diet, and once very common in the Orient due to the heavy consumption of polished rice.

beta decay. A particular type of radioactive disintegration.

Big-Bang theory. In astronomy, a theory that the universe began with a gigantic explosion from super-condensed matter.

bile, n. A yellowish-green secretion of the liver that aids in the digestion of fats.

biofeedback method. Any one of several techniques that use electrical devices for converting biological pressures, caused, for example, by mental stress, so that they can be heard as rhythmical beats or viewed as waves on a screen.

biological clock. An intrinsic mechanism, thought to be located in the brain, by which an individual can measure time.

biological warfare. Warfare involving the use of disease germs and herbicides.

biomass, n. The total quantity of living organisms, plant or animal, of one species per unit of area.

biorhythm, n. A cyclic pattern or variation in one or more parts of the body.

biotin, n. A vitamin B component associated with growth and found abundantly in yeast and liver.

bipolar disorder. (See: manic depression.)

black body. In physics, any object that absorbs light totally, that is, without reflection.

black box. Usually, any specifically designed, electrical control mechanism.

black hole. An as yet unobserved area in outer space having super-gravitational attractive forces, and whose existence is a theoretical consequence of the equations of general relativity.

black lung disease. A deteriorated lung condition due to excessive inhalation of coal dust.

blastula, n. An early form of an embryo, which consists of a single layer of cells that has formed into a spherical shape.

blood plasma. The fluid part of the blood.

blood pressure. The pressure exerted by blood on the walls of blood vessels, especially arteries.

blood type. A blood classification determined by the nature of the red cells; successful transfusion is possible only between matched types.

blue whale. A marine mammal that is the largest of all known mammals.

bone marrow. The soft tissue that fills the central, cavity sections of many bones and manufactures blood cells.

botany, n. The study of the structure and function of plants.

botulism, n. A form of severe poisoning from a bacillus toxin, which may infect food that has been canned, jarred, or preserved improperly.

brain damage. Any injury to the brain that results in a mental or physical disability.

brain waves. Rhythmic electrical fluctuations due to the flow of current in the brain.

breeder reactor. A nuclear reactor that yields new, fissionable material as a byproduct while generating energy.

bronchi, n. The two main tubular organs connecting the lungs to the windpipe.

Brownian motion. The very minute motions of atoms and molecules in a liquid.

bubble chamber. A device that contains the vapor bubbles of a super-heated liquid, in which one can follow the path of a moving subatomic particle.

bubonic plague. (See: plague.)

bypass surgery. Replacement of one or more blood vessels that bring blood to the heart muscle so that it can pump and function properly.

C

Cesarean section. Surgical delivery of a fetus through an abdominal wall.

caffeine, n. A stimulant alkaloid present in tea, coffee, and kola.

calcify, v. To make or become stony or calcareous.

calorie, n. In general, a unit of measurement of heat; in particular, a measure of the energy or heat-producing value of a particular food.

cancer, n. Any malignant, uncontrolled cellular growth.

capillary, n. The smallest type of blood vessel in the body, from which nutrients and oxygen pass into cells and into which the cells deposit metabolic wastes and carbon dioxide.

capillary action. The rise and fall of a liquid in very thin tubes or in porous media, due to forces interacting between the solid and liquid interfaces.

carbohydrate, n. Any one of a group of substances composed of carbon, hydrogen, and oxygen, including the sugars and starches; carbohydrates are produced by all green plants by the process of photosynthesis.

carbon dioxide. A gas produced by animals in respiration and by the combustion of carbon-containing materials; plants use carbon dioxide to produce carbohydrates by photosynthesis and, simultaneously, create oxygen as a byproduct.

carbon monoxide. A highly poisonous gas produced by the incomplete combustion of carbon, a most common source of which is the standard automobile engine.

carcinogen, n. Any cancer-producing substance.

cardiac arrest. Failure of the heart to function.
cardiac arrhythmia. Irregularity in the time or force of the heartbeat.
cardiac infarction. A sharply limited region of hemorrhage or cell death in the heart due to an obstruction of the blood circulation in, or to, the heart muscle.
cardio-pulmonary resuscitation (CPR). A first-aid technique in which one blows air into a nonbreathing person to provide oxygen and presses down on the chest bone rhythmically to keep the heart pumping.
cardiovascular bypass. (See: bypass surgery.)
cardiovascular disease. Any disease or abnormal condition of the heart and the blood vessels.
cardiovascular system. The heart and all the blood vessels in the body.
caries, n. Tooth decay.
carotene, n. A pigment in plants and animals that imparts a yellow, orange, or red color.
catalyze, v. To accelerate.
cataract, n. An opaque tissue formation in or over the lens of the eye.
catarrh, n. Inflammation of mucous membranes accompanied by a marked increase in mucous flow, especially from the nose.
cathode ray tube. A vacuum tube that allows an electric discharge from one end, called the cathode, to the other, called the anode.
CAT scan. A computer-aided technique for using a series of X rays to reveal internal organs in cross sections; computer-aided tomography; computer axial tomography.
cell, n. In biology, the smallest possible unit of living matter, identified by a centrally located, concentrated mass called the nucleus, and by a clearly delineated boundary membrane called the cell wall.
cellulose, n. A complex of sugars that is the basic component of the cell walls of plants.
central nervous system. The brain and the spinal cord.
centrifugal force. A force radially outward from a central point.

centrifuge, n. A rotating device used to separate out materials of different densities.

centripetal force. A force radially inward toward a central point.

cerebellum, n. A large posterior section of the brain that coordinates the muscles and maintains body equilibrium.

cerebral palsy. A disability due to brain damage, characterized by muscular incoordination, spastic paralysis, and speech disorder.

cerebrum, n. A large frontal part of the brain, associated primarily with thinking and sensory processes.

chain reaction. The self-sustained, continued release of energy by the splitting of atomic nuclei, which follows from a first such reaction.

charged particle. Any particle that has a positive or a negative charge.

chemical analysis. The determination of the chemical components of a given substance.

chemical bond. The union of two or more atoms into a single unit, called a molecule.

chemical warfare. Warfare involving chemicals different from explosives, like the use of poison gas.

chemotaxis, n. Movement of an organism in response to chemical stimulation.

chemotherapy (chemo), n. Treatment of disease with drugs, associated often with various cancer treatments.

chiropractic, n. A treatment to cure a disease or to relieve pain by manipulation of the spine.

chlorophyll, n. The green matter of plants, which converts carbon dioxide and water under the stimulus of light into natural food and oxygen.

chloroplast, n. A small, structural unit in a leaf that contains chlorophyll.

cholera, n. An acute, bacterial disease characterized by vomiting and severe, dehydrating diarrhea.

cholesterol, n. A substance consisting, chiefly, of fatty acids.

chromosome, n. The threadlike part of the cell nucleus that carries hereditary information in the form of genes.

cilia, n. A minute, hairlike structure.

circadian rhythms. Physiological cycles, approximately twenty-four hours in length, which regulate a variety of activities in plants and animals.

cirrhosis, n. A degenerative disease of the liver.

citric acid cycle. A theoretical sequence of cell reactions, which describes steps by which cells produce energy; also called the Krebs cycle.

clone, n. An exact reproduction, without genetic change, of a plant or animal.

cloud chamber. A device containing saturated water vapor that reveals the path of a charged particle by a trail of visible droplets.

cloud seeding. An aerial process of scattering a chemical substance into clouds to cause rain.

coagulate, v. To clot.

coal-tar, n. A thick, opaque liquid distilled from bituminous coal.

cocaine, n. A narcotic obtained from the dried leaves of the shrub E. coca.

cochlea, n. A bony structure of the inner ear which is shaped like a snail shell and contains the end organ of hearing.

cocoon, n. A protective covering in which many insects spend the pupa stage of development.

coitus, n. Sexual intercourse.

colitis, n. Inflammation of the mucous membrane of the large intestine.

colostomy, n. A surgically created exit for waste material from the bowel, which voids the need for direct, rectal elimination.

comatose, adj. Lethargic, as if in a coma.

comet, n. A luminous heavenly body with a long nebulous tail, and known in some cases to be in orbit around the Sun.

computer, n. A high-speed arithmetic processor that can be programmed to operate logically and that has a memory unit for storage and retrieval of programs and data.

condenser, n. Generally, any device for changing vapor to liquid; in physics, a device for storage of electricity.

conger, n. A large sea eel.

conjunctivitis, n. Any of several infections of the eye.
continental drift. The relatively slow, primarily horizontal motion exhibited by large land masses.
continental shelf. That portion of a large land mass that extends gradually into the ocean to a point where it drops suddenly and the water becomes extremely deep.
contraception, n. The prevention of conception.
convective current. Usually, a streaming or circulatory motion in a gas or a liquid due to heating.
Coriolis force. A deflecting force on a body due to the Earth's rotation.
cornea, n. The transparent, outer coating of the eyeball.
corona, n. Any crownlike structure.
coronary artery. An artery that supplies blood to the muscle cells of the heart, which is thereby provided the energy to pump blood to all other cells of the body.
coronary bypass. Surgical replacement of one or more coronary arteries; bypass surgery.
coronary occlusion. A blockage in a coronary artery that prevents the muscle cells of the heart from receiving adequate oxygen and nutriments.
corpuscle, n. A red or a white blood cell.
cortex, n. The outer part of an organ, especially that of the brain.
cortisone, n. An adrenal gland compound, also produced synthetically, and used extensively for the relief of pain.
cosmic ray. Any stream of high-energy particles that enters the Earth's atmosphere from outer space.
cosmology, n. The study of the universe.
cosmos, n. The universe.
CPR. (See: cardio-pulmonary resuscitation.)
Crab Nebula. A dense, glowing gas between the stars of the constellation Taurus.
cretin, n. An individual with a congenital thyroid disorder, characterized by small height, deformities, and low mental capability.
cryogenics, n. The study of very low temperature refrigeration.
crystal, n. A solid formation characterized by identical substruc-

tural, geometrical configurations, as in a quartz or a snow crystal.

CT-scan. (See: CAT scan.)

cumulus cloud. A massive cloud formation characterized by a flat base and rounded formations piled up like a mountain.

curare, n. A vine extract used both as an antispasmodic and an arrow poison.

cybernetics, n. The study of electrical systems as analogues of the human brain and nervous system.

cyclone, n. A storm like a tornado.

cyclotron, n. A special type of particle accelerator.

cystic fibrosis. A glandular disease in which fibrous mucus is produced and forms blockages in the body.

cystitis, n. An inflammation of the bladder, often accompanied by frequent and painful urination.

cytology, n. The study of the structure and function of biological cells.

cytoplasm, n. The protoplasm of the cell, excluding the nucleus.

D

Darwinism, n. The theory of biological evolution.

dementia, n. Progressive loss of intellectual function, including impairment of memory, judgment, and thought processes.

dengue, n. A tropical disease, primarily, which is epidemic and characterized by fever, skin eruption, and severe bodily pains.

deoxyribonucleic acid (DNA). A constituent of chromosomes in nuclei that contains genetic data in a coded form.

depression, n. Clinical depression is a mental state characterized by intense sadness, crying, an inability to concentrate, loss of appetite, sleeplessness, retreat from social contacts, and, often, attempted suicide.

dermatology, n. The study of the structure, function, and diseases of the skin.

diabetes, n. A disease marked by the failure of the pancreas to secrete insulin, preventing the body from utilizing properly all dietary carbohydrates.

dialyze, v. To pass through a suitable membrane for filtering.

diffraction, n. The bending of a wave when it passes from one medium, like air, into another, like water.

digitalis, n. The dried leaf of the common foxglove, used as a cardiac stimulant and diuretic.

diphtheria, n. A contagious, bacterial disease in which air passages become coated with a membrane formed by a fibrous discharge.

direct current (DC). A type of electric current characterized by electrons moving in a fixed direction.

diuretic, n. Any drug that produces an increased flow of urine.

DNA. (See: deoxyribonucleic acid.)

DNA evidence. *Two* sets of DNA fragments, obtained under controlled conditions. Almost identical sets from the scene of a crime and from a suspect in the crime are used to support guilt. Unidentical sets are used to support innocence.

dopamine, n. A substance that functions as a transmitter for nerve cells in the brain.

double helix. The two identical DNA strands of genetic material found in each cell nucleus and entwined in a helical form.

double star. Two stars situated relatively close to each other and rotating around a common axis.

Down's syndrome. The physical traits associated with a birth-defective child usually called a mongoloid.

drosophila, n. A class of flies, including the common fruit fly, used extensively in recent genetic experiments.

drug addiction. The mental and physical habitual need for a drug.

Dutch elm disease. A fungus disease of elm trees characterized by yellowing of leaves, defoliation, and death.

dysentery, n. An inflammation of the large intestine, usually bacterial in nature, accompanied by pain, a need to evacuate the bowels, and a discharge of blood and mucus.

dyslexia, n. A disturbance of the ability to read.

E

eclipse, n. The obstruction of light by one celestial body from another.

E coli. A particular intestinal bacterium, Escherichia coli, associated with genetic experimentation and outbreaks of food poisoning.

ecosystem, n. An ecological system, that is, one in which a variety of organisms live in a natural balance.

ectoderm, n. The outermost layer of germ cells.

eddy, n. Any whirlpool, large or small.

EEG. (See: electroencephalogram.)

effluvium, n. In biology, a noxious emanation.

EKG. (See: electrocardiogram.)

elastin, n. The chief component of elastic fiber.

electricity, n. The flow of electrons.

electrocardiogram (EKG), n. A graphical record of electrical potential changes that occur during the heartbeat.

electroencephalogram (EEG), n. A graphical record of electrical brain wave patterns.

electron, n. A fundamental, relatively light atomic particle with a negative electrical charge.

electron microscope. A microscope of exceptional sensitivity that uses the character of electron beams for magnification.

embryology, n. In mammalian biology, the study of the development of an organism while it is still in its mother's body.

emetic, n. Any substance that induces vomiting.

emphysema, n. A disease of the lungs characterized by scarred tissue and excessive fluid accumulation.

encephalitis, n. A form of sleeping sickness.

endocrine system. Those glands whose secretions pass directly into the blood, like the pituitary and the thyroid.

endoderm, n. The innermost layer of germ cells.

endorphin, n. A glandular secretion with the properties and effects of morphine.

endotoxin, n. Any substance found inside a bacterial cell that acts as an antigen.

energy, n. In physics, any quantity that has the ability to do work, like heat and electricity.

enteritis, n. Inflammation of the small intestine.

entomology, n. The study of insects.

entropy, n. A measure of the degree of disorganization of a system.

enucleation, n. The removal of a nucleus.

enzyme, n. In biology, any substance that accelerates specific chemical transformation, like those required for the digestion of food.

epidemiology, n. The study of the treatment of epidemics.

epidermis, n. Usually, the outer layer of an animal's skin.

epilepsy,n. A neurological disorder marked by convulsions, loss of consciousness, and frothing at the mouth.

equinox, n. A time of the year when the Sun's position crosses the equatorial plane and the length of time of both day and night are the same.

erythrocyte, n. A red blood cell.

esophagus, n. The tube connecting the back of the mouth with the stomach; the gullet.

ESP. (See: extrasensory perception.)

estrogen, n. A female hormone that influences the reproductive process.

ether, n. In physics, a medium assumed by some to permeate all space, which permits the transmission of electromagnetic waves.

euthanasia, n. The practice of putting to death painlessly one who is suffering from an incurable disease or disability.

evolution, n. In biology, a theory that all plants and animals have evolved from simpler forms and have reached their present structures by the processes of environmental adaptation and survival of the fittest.

exotoxin, n. Any poison excreted by a microorganism.

extrasensory perception (ESP). A possible mental capability to sense things that are not present, as in clairvoyance and telepathy.

F

fallopian tube. A duct through which an egg cell travels from an ovary to the uterus.

fault, n. In geology, a fracture in the Earth's crust.

fauna, n. Animal life.

febrile, adj. Feverish.

feces, n. Excrement.

fermentation, n. A chemical change with effervescence, as that produced by yeast.

fermion, n. A subatomic particle with a particular type of spin and momentum.

ferrites, n. Compounds of iron.

fetus, n. An animal embryo in the womb.

fibril, n. A small fiber, thread, or hair.

fibrillation, n. A form of heart disease in which groups of muscle fibers of the heart beat independently, often rapidly, without rhythm.

fibrin, n. A fibrous protein formed in blood coagulation.

fireball, n. A luminous meteor, often called a "shooting star."

fission, n. In physics, the splitting of an atomic nucleus; in biology, the process of cellular asexual reproduction.

fissure, n. A narrow opening; a crack.

fistula, n. An abnormal, pipelike passage in an abscess or a hollow organ.

flagellum, n. A whiplike appendage that serves as a swimming organ in minute organisms.

flora, n. Plant life.

fluoridate, v. To combine with a chemical fluoride, as in the case of drinking water.

fluoride, n. A chemical compound containing fluorine.

fluorine, n. A corrosive gas of the chlorine family.

folic acid. A particular B-complex vitamin associated with growth.

food preservative. Any chemical additive to food, like sodium benzoate, which appreciably decreases the rate of decay.

foot-and-mouth disease. An acute, contagious disease of cattle, sheep, and swine characterized by ulcers about the hoofs and in the mouth.

fossil, n. Any trace of an animal or plant of past geological ages that has been preserved in the earth's crust.

fungus, n. A plant that reproduces by means of spores, not seeds, and is devoid of chlorophyll, like a mushroom.

fusion, n. The union of two atoms in such a fashion that their individual nuclei become a single nucleus.

G

galaxy, n. An assembly of stars, all in rotation around a central core, like the Milky Way galaxy.

gallstone, n. A pebblelike formation in or adjacent to the gall bladder.

galvanic cell. A device for producing electricity by chemical means; a battery.

gamete, n. Any generative or reproductive cell.

gamma rays. Radiation similar to X rays, but of a shorter wave length.

gastrointestinal, adj. Pertaining to the stomach and the intestines.

Geiger counter. A sensitive, electrical device for detecting radioactive materials and waves.

gemstones, n. Minerals of high value due to beauty, durability, or rarity, including diamonds, rubies, and emeralds.

gene, n. The fundamental biological unit for transmitting a hereditary characteristic.

generic name. The actual or universal name of a substance, as opposed to a marketing, business, or trade name.

gene splicing. A biological engineering technique of replacing genes or joining genes that have come from a different source cell.

genetic code. A theory that explains how the basic DNA units, that is, adenine, guanine, cytosine, and thymine, provide the code for the basic amino acids.

genetic disease. Any disease or abnormal condition due to a defect in the character of at least one gene.

genetic engineering. That branch of engineering devoted to the development of tools and techniques for the manipulation of genes.

geodesy, n. The measurement of large portions of the Earth's surface.

geology, n. The study of the structure and composition of planets and moons, especially of Earth.

geomagnetism, n. The natural magnetism of planets and moons.

geothermal, adj. Pertaining to the natural heat of Earth's interior.

geotropism, n. The involuntary reaction of plants and animals to the force of gravity.

geriatrics, n. The medical study of old age and its diseases.

germ tissue. Cells from which an embryo develops.

gerontology, n. The scientific study of the phenomena of old age.

gill, n. An organ for animal respiration in water, enabling fish, for example, to extract oxygen from water and return carbon dioxide.

glacier, n. A moving field or body of ice.

glaucoma, n. A hardening of the eyeball that results in impairment of vision or blindness.

global warming. A theory that the burning of fossil fuels, like oil and gasoline, contributes substantially to a warming trend of the atmosphere, which could disrupt weather patterns, ecosystems, and agriculture worldwide.

globulin, n. A general term for a class of proteins that do not dissolve in water, like those from beans and nuts.

glucose, n. A chemical form of sugar.

gluon, n. A theoretical subatomic substance that holds things together.

glycerine, n. A sweet, colorless, sticky, and viscous liquid derived from fats and oils.

goiter, n. An enlargement of the thyroid gland due to a shortage of dietary iodine.

gonad, n. An ovary or a testicle.

gonorrhea, n. A contagious bacterial disease affecting the genital and urinary tracts.

gout, n. A metabolic disease marked by pain and swelling of the joints.

gravitation, n. The force of attraction between any two bodies in the universe.

gravity, n. The force of attraction between Earth and a body on a or near Earth's surface.

Great Red Spot. A prominent red spot on the surface of Jupiter.

greenhouse effect. The development of a layer of carbon dioxide gas in the atmosphere, which traps heat and warms the globe.

green revolution. The vast increase in crop production due to the application of modern scientific discoveries.

gynecology, n. The branch of medicine that treats women, their hygiene, and their diseases.

H

hagfish, n. An eellike fish.

H-bomb. (See: hydrogen bomb.)

Halley's comet. A particular comet that orbits the Sun periodically, thus providing proof that certain comets, at least, are constituents of our solar system.

hallucinogen, n. Any agent that induces hallucinatory-type perception.

hazardous waste. Any radioactive or chemically poisonous waste product.

heart attack. (See: cardiac infarction.)

heart disease. (See: cardiovascular disease.)

heavy water. A form of water in which the hydrogen atom is replaced by an isotope of hydrogen.

heliotropism, n. The tendency of flowers and leaves to face toward the Sun.

helium, n. Next to hydrogen, the lightest of gases; the second element in the table of elements.

hematite, n. A red mineral that is important as an ore of iron.

hematology, n. The study of the blood.

hemoglobin, n. A protein in red blood cells that carries oxygen to the tissues.

hemophilia, n. A genetic disease in which blood does not clot.

hepatitis, n. A disease of the liver, one form of which is called "yellow jaundice."

herb, n. Any plant with a soft, succulent stem, which withers away after flowering and is used for medicinal purposes or for its flavor.

herbicide, n. Any chemical that destroys plants.

herbivore, n. Any animal that eats only plants.

hermaphrodite, n. Any individual with both male and female reproductive organs.

Heroin, n. A trademark of the white, crystalline narcotic diacetyl-morphine.

herpes, n. A disease of the skin and mucus membranes, of a viral nature, and often recurring.

high-energy physics (See: particle physics.)

histamine, n. A substance released by the tissues during conditions of inflammation or allergy.

HIV (human immunodeficiency virus). Human immunodeficiency virus is transmitted through contact with blood, semen, or vaginal fluids of an infected person and is associated with AIDS.

holography, n. A method of producing three-dimensional images by use of light waves that are in step.

hominid, n. A modern or contemporary human.

homo sapiens. Man, regarded as a biological species.

hoof-and-mouth disease. (See: foot-and-mouth disease.)

hormone, n. Any internal glandular secretion that promotes activity in cells.

hybrid, n. Any crossbreed, animal or plant.

hydrate, n. Any compound substance that contains water as one constituent.

hydrocarbon, n. A general name for bitumens, resins, and fats that are composed of carbon and hydrogen.

hydrochloric acid. A hydrogen and chlorine compound with a variety of commercial uses; also found naturally in stomach juices.

hydrogen, n. The lightest gas, the smallest atom, and the first element in the table of elements.

hydrogen bomb. A fusion weapon that is detonated by an atomic bomb to create the heat necessary for hydrogen fusion; a thermonuclear bomb.

hydrogen peroxide. A compound that serves as a bleaching agent and an antiseptic.

hydrolysis, n. A chemical decomposition process that utilizes water.

hydroxide, n. A complex of molecules, each containing one hydrogen and one oxygen atom.

hypertension, n. High blood pressure.

hypnosis, n. A poorly understood state of high concentration characterized physically by a state of sleep and mentally by direct accessibility to subconscious activities of the brain.

hysteresis, n. Any lag in an effect due to a change of force.

I

Ice Age. A period in Earth evolution when large portions of its surface were covered with glaciers.

immunological system. Those bodily organs that contribute to the body's ability to resist disease-infecting organisms.

inertia, n. The disposition of an object at rest to remain at rest and of an object in motion to remain in motion.

influenza, n. A group of viral, communicable diseases marked by throat and bronchial inflammation and pains in the muscles and joints.

infra red. Pertaining to electromagnetic waves whose lengths are slightly greater than those of red light.

insect, n. Any animal with three body sections, three pairs of legs, and no back bone.

insulin, n. A pancreatic substance necessary for the body to break down sugars into usable food forms.

integrated circuit. An electrical circuit with several independent parts which have been combined to do a special job.

intelligence, n. Innate mental ability.

interferon, n. An antiviral protein produced by human cells that have been invaded by viruses.

ion, n. An atom that is electrically charged by virtue of having gained or lost at least one electron.

ionosphere, n. An atmospheric region of electrically charged air, which begins at an altitude of about twenty-five miles.

isobar, n. A line of constant pressure on a weather map.

isometric, adj. Of equal measure.

isotherm, n. A line of constant temperature on a weather map.

isotope, n. Any of two or more forms of the same element, distinguishable primarily by different masses, due to having a different number of neutrons in the nucleus.

J

jet stream. Strong air currents that flow from west to east at over thirty thousand feet and affect the development of high and low pressure systems.

jugular vein. A major vein in the neck that returns blood from the head to the heart.

K

Kreb's cycle. (See: citric acid cycle.)

L

lacrimal gland. A tear gland.

lactic acid. An acid commonly found in fermented milk products, like cheese and buttermilk, but also present in the blood.

larva, n. The immature, wingless, often worm-like form in which certain insects hatch from the egg.

laser beam. A beam of super-concentrated light.

Lassa fever. A serious viral disease of African origin characterized by high fever, body rash, ulcers in the throat, and swelling about the face and the neck.

latex, n. A milky plant fluid from which rubber is derived.

laughing gas. (See nitrous oxide.)

lectin, n. Any of several plant proteins that induce red blood cell clumping.

Legionnaire's disease. One of several serious types of bacterial pneumonia.

legume, n. The fruit or seed of a pod-bearing plant, which is used for food, as, for example, peas.

lepidopteran, n. A moth or butterfly.

lepton, n. Any member of the class of lightest subatomic particles.

lesion, n. An injury; a morbid change in the structure of an organ.

leucocyte, n. A white blood cell.

leukemia, n. Cancer of the blood.

lice, n. The plural of louse.

lichens. n. A plant group, similar to moss, which grows on rocks.

lignin, n. A substance like cellulose in woody tissue.

limnology, n. The study of the physics, chemistry, and biology of lakes.

lipids, n. A group of substances comprising the fats, characterized by insolubility in water and a greasy feeling.

lobotomy, n. The severing of nerve fibers in the frontal brain lobe supposedly for the relief of mental disorders.

lockjaw, (See: tetanus.)

louse, n. A small, flat wingless insect, which is parasitic on warm-blooded animals.

LSD. (See: lysergic acid diethylamide.)

lugworm, n. Any one of a common class of marine worms that are hermaphroditic.

lymph, n. A colorless, coagulable fluid consisting mainly of blood plasma and colorless blood cells.

lymphocyte, n. A white colorless blood cell derived from lymphatic tissues.

Lysenko theory. A biological theory that maintains that learned or acquired traits can be transmitted genetically to succeeding generations.

lysergic acid diethylamide (LSD). An acid that is one of a variety of hallucinogenic substances.

M

Mach number. Usually, the ratio of the speed of a plane or rocket to the speed of sound in air.

magma, n. Molten rock.

magnetic field. Usually, the area of magnetic influence around a magnet.

magnetism, n. The property of the molecules of certain substances to attract iron.

malnutrition, n. Generally, an unhealthy bodily condition due to faulty diet or lack of food.

mammal, n. Any vertebrate family in which the female nourishes its young naturally with milk.

manic depression (bipolar disorder). A form of mental illness characterized by alternating and extreme feelings of elation and depression.

marijuana, n. A hemp cannabis sativa; also its dried leaves and flowers.

marsupial, n. Any animal that has a pouch for carrying its young, like a kangaroo.

maser, n. A device that produces an intense beam of microwave radiation of wave length shorter than that of a laser.

mass, n. The quantity of a substance, often measured by its weight.

mastectomy, n. Surgical removal of a breast.

medulla, n. The posterior part of the brain, just above the spinal cord.

melanin, n. Any of the darker skin pigments, present in all animal skins, to a greater or lesser degree, except those of albinos.

membrane, n. A thin, soft, pliable sheet of human tissue.

menopause, n. The period of natural cessation of ovulation.

menstruation, n. The uteral discharge of bloody fluid and other material, which has been prepared by the body for the growth of an embryo.

mental illness. Any disorder of the brain that affects a person's judgment and behavior and renders the person, to some degree, dysfunctional.

mesoderm, n. The middle layer of germ cells.

meson, n. A type of subatomic particle between the very light leptons and the relatively heavy baryons.

metabolism, n. In general, that process in all animal cells by which the cell converts food to energy and removes the associated waste products.

metallurgy, n. The study of the extraction of metals from ores and their refinement for use.

metamorphosis, n. A change from one form into another.

meteorology, n. The study of the atmosphere, especially of its moisture, winds, and heat variations.

meteors, n. Rocklike, celestial bodies, both small and large, which may enter Earth's atmosphere with great speed, light, and heat; often called shooting stars.

methaqualone, n. A drug that induces sedation.

methyl alcohol. A poisonous form of alcohol derived from distillation of wood; also called wood alcohol.

metric system. A measurement system using decimal units, the unit of length being the centimeter and of weight being the gram.

microbe, n. A very minute organism not visible to the naked eye; a microorganism; popularly, a pathogenic bacterium.

microsurgery, n. Surgery on a minute part of the body, like a nerve, using special tools, like laser beams.

microwave, n. Any electromagnetic wave of very short wave length.

Milky Way galaxy. The galaxy of Earth's Sun, visible at night as a broad, milky-colored band of stars across the sky.

mitochondria, n. Granular or rod-shaped parts of a living cell, which aid in the process of metabolism.

mitosis, n. Cell reproduction by division into two distinct new cells.

modulator, n. Often, a device for changing the frequency of electrical waves.

molecule, n. A unit of matter that consists of two or more bonded atoms.

mollusk, n. A shellfish that does not have a horny shell; for example, a snail, mussel, or oyster, but not a lobster or a crab.

monosaccharide, n. A simple sugar.

moon paradox (illusion). The apparent enlargement of the moon's appearance when it is near the horizon.

morphine, n. The white, crystalline, narcotic base of opium.

morphogenesis, n. The evolution or development of biological form or structure.

morphology, n. In biology, the study of the form and structure of animals and plants.

mucous membrane. The lining membrane of those cavities of the body that communicate directly or indirectly with the exterior.

mucus, n. A viscous, slippery secretion produced by mucous membranes, which it moistens and protects.

mulch, n. Vegetable matter applied to the surface of the soil around plants to enhance growth.

multiple sclerosis. A neurological disease characterized by paralysis and muscle tremors.

muon, n. A subatomic particle similar to an electron but about 200 times heavier.

mushroom, n. Any of several fleshy, edible fungi.

musk, n. A strong scented substance from a sac near the abdomen of the male musk deer, which is used in many perfumes.

mutation, n. A sudden and exceptional variation in offspring due to genetic changes.

myoblast, n. A cell that gives rise to muscle tissue.

myocardial infarction (See: cardiac infarction.)

myoglobin, n. A muscle protein associated with the storage and use of oxygen.

myosin, n. The most abundant protein in skeletal tissues, which acts with actin in the contraction and relaxation of muscles.

N

nanotechnology, n. A new branch of technology in which exceptionally small mechanical and electrical devices are constructed by manipulation with atoms and molecules.

narcolepsy, n. A neurological condition characterized by periods of lengthy and deep sleep.

narcotic, n. Any drug that in moderate amounts relieves pain, induces euphoria and sleep, and results in physical dependency with time, but that in large amounts induces stupor, coma, or convulsions.

natural selection. The evolutionary process by which plants and animals that are best adapted to the natural environment grow and reproduce, while those that are poorly adapted die out.

nebula, n. A dense, glowing gas between stars.

necrotic tissue. Dead tissue.

negative ion. An atom that has acquired at least one additional electron and is no longer electrically neutral, but has a negative charge.

neoplasm, n. A tumorous form of growth.

nephritis, n. Inflammation of a kidney.

nephrology, n. The study of kidneys.

nervous system. In humans, the brain, the spinal cord, and all the nerve cells in the body.

neuralgia, n. Acute pain in one or more nerves.

neurology, n. The study of the nervous system.

neutrino, n. The most penetrating of all subatomic particles, traveling at the speed of light, having no charge, and having a mass so small that it has not been measured precisely.

neutron, n. An uncharged particle, of slightly greater mass than the proton, which can be found in all atomic nuclei except that of hydrogen.

neutron star. An X ray emitting star composed largely of neutrons, which are bound more strongly than ordinary nuclear bonding forces.

nicotine, n. A poisonous alkaloid that is the active principle of tobacco.

nitrogen, n. An element that is found abundantly in the air and in the soil, and that is a basic component of living tissue.

nitrogen cycle. Aided by bacterial actions, the sequential passage of nitrogen from air, into the soil, into plants, into animals, and back into the air and the soil.

nitrous oxide. An anesthetic used in dentistry, which produces giddiness before insensibility to pain; laughing gas.

Northern Lights (See: aurora borealis.)

nuclear energy. Energy produced by nuclear fission or fusion.

nuclear fission. The division or splitting of the nucleus of an atom.

nuclear fusion. The uniting of the nuclei of two atoms into a single nucleus.

nuclear reactor. A device for creating energy by splitting atomic nuclei.

nucleation, n. The initial process of the formation of a crystal from a liquid or a gas.

nucleic acid. Either one of DNA or RNA.

nucleon, n. Any particle in an atomic nucleus, including protons and neutrons.

nucleus, n. In an atom, the central portion, containing most of the atom's mass and having a positive charge; in a biological cell, the central, dense portion that contains DNA and a variety of other components related to the cell's physiology and functioning.

nutrient, n. Any nourishing substance.

nutrition, n. The process by which animals and plants ingest and use food substances.

O

oceanography, n. The study of the geography of oceans and of phenomena related to oceans.

oil shale. Bituminous coal that contains oil.

oleoresin, n. A natural plant resin in an essential oil.

olfaction, n. The sense of smell.

ophthalmologist, n. A medical doctor who specializes in abnormalities and diseases of the eye, and who prescribes corrective optical lenses.

opiate, n. Any medicine derived from opium.

optician, n. Primarily, one who makes optical glasses.

optic nerve. That particular nerve which connects the eye and the optic center of the brain.

optometrist, n. Primarily, one who prescribes correctional optic lenses.

oral contraceptive. A pill or other substance taken through the mouth, which prevents conception.

organelle, n. Any of several specialized structures within a cell.

organic, adj. In biology, pertaining to, or derived from living organisms; in chemistry, pertaining to or designating that branch which deals with the compounds of carbon.

ornithology, n. The study of birds.

oscilloscope, n. A device for showing changes in current by means of a wavy line on a fluorescent screen made by electron beam deflection.

osteopathic medicine. A medical system based on the theory that disease is due, primarily, to mechanical derangement of tissues, placing emphasis on restoration of the structural integrity of the body.

ovary, n. A female organ that produces and discharges eggs.

ovulation, n. Production and discharge of eggs from an ovary.

ovum, n. An egg cell.

oxidation, n. Combination with oxygen.

oxygen, n. The most abundant element on the Earth's surface and which is indispensable in animal respiration and metabolism.

ozone, n. A form of oxygen derived by means of electrical discharge.

ozone layer, n. A region of the upper atmosphere that contains a high concentration of ozone and is associated with the filtering of cosmic rays.

ozone hole. An opening in the ozone layer above Antarctica.

P

pacemaker, n. An electrical device for controlling the rate of the heart beat.

palate, n. The roof of the mouth.

paleontology, n. The study of fossils and of life in past geological ages.

palpate, v. To examine a part of the body by feeling with the fingers.

pancreas, n. A large gland that secretes enzymes into the intestine to aid in the digestion of food.

pandemic, adj. Affecting a majority of people; epidemic.
paramecium, n. A type of protozoan covered completely with fine hairlike filaments, which are used for propulsion and gathering of food.
parapsychology, n. The psychology of extrasensory perception.
parasite, n. An animal or plant living in or on another living organism and feeding on that organism.
paregoric, n. A camphorated tincture of opium.
parity, n. In physics, loosely, transformation into a mirror image.
Parkinson's disease. A disease of the central nervous system characterized by physical shaking, tremors, loss of balance, and poor articulation.
particle accelerator (See: accelerator.)
particle physics. The physics of subatomic particles.
pathogenic, adj. Disease producing.
pectoral, adj. Situated on the breast or chest.
penicillin, n. A broadly applicable, antibacterial substance, which is produced synthetically and occurs naturally in green mold.
penis, n. The male organ for sexual intercourse and urination.
pepsin, n. A stomach enzyme useful for the digestion of food.
pH. In chemistry, a number on the scale 0–14, in which less than 7 indicates a level of increasing acidity, more than 7 indicates a level of increasing alkalinity, and 7 indicates neutrality.
phage, n. (See bacteriophage.)
pharmacology, n. The study of drugs and medicines.
phlebitis, n. (See: thrombophlebitis.)
phlogiston, n. A hypothetical material once thought to be the substance of heat energy.
phonon, n. An energy packet associated with the vibration of atoms about their rest positions and used in the modeling and analysis of sound waves.
phosphate, n. A compound derived from phosphoric acid.
phosphene, n. A sensation of seeing light when none is present, often caused by application of pressure to the eyeball.
phospholipid, n. A type of body fat that is high in phosphorous.
phosphoric acid. An oxygen acid of phosphorous.

phosphorous, n. An active element, similar in its properties to nitrogen, characterized usually by its disagreeable smell.

photon, n. The smallest possible unit of light energy.

photosynthesis, n. The process by which green plants convert light and carbon dioxide into carbohydrates.

phototropism, n. A natural motion, usually of plants, toward light.

physiology, n. Generally, the study of the activities and functions of living organisms and of particular organs.

pigmentation, n. Color in the skin, eyes, or hair, due primarily to melanin.

pituitary gland. A small gland near the brain that promotes skeletal growth.

placebo, n. Often, an inert medication, of no intrinsic physical value, prescribed to satisfy a patient's mental need for medication.

placenta, n. The tissue structure connected by the umbilical cord to the fetus, which nourishes the fetus in the uterus; often identified as the afterbirth.

plague, n. A serious bacterial disease of rats, transmitted to humans by fleas, which can affect the lymph glands (bubonic plague), the lungs (pneumonic plague), or the blood stream (septicemic plague), and which was associated with past epidemics of such magnitude that it was called the Black Death.

plankton, n. Minute animals and plants that live in water but whose concentrations can be so large as to provide food for whales.

plasma, n. (See: blood plasma.)

plastics, n. A group of organic, synthetic, and processed materials, which can be molded or cast.

platelet, n. A small, colorless, round body, without a nucleus, which is important in the clotting of blood.

plate tectonics. A geological theory that has Earth's surface divided into six major and several minor rocky plates, which move in reaction to the earth's internal forces.

Pleiades, n. An open star cluster in the constellation Taurus.

plexus, n. A network of interlacing blood vessels or nerves.

plutonium, n. A radioactive element that can be used as a nuclear fuel.

pneumococcus,n. A bacterium that causes lobar pneumonia.

pneumonia, n. A disease characterized by inflammation and fluid accumulation in one or both lungs.

poliomyelitis, n. A viral disease, usually of children, which affects the spinal cord and often induces permanent paralysis; also called infantile paralysis.

pollen, n. Microspores of seed plants, almost dustlike in character.

polygraph, n. An electronic device for recording several bodily sensations simultaneously; also called a lie detector.

polymer, n. A substance that consists of many molecules of the same kind, arranged in a chainlike fashion.

polyp, n. A swollen, hardened mass of mucous membrane.

polypeptide, n. Any of several amino acid, nonprotein compounds.

polysaccharide, n. A compound of two or more simple sugars.

poppy, n. A plant characterized by a milky juice and attractive red flowers, one type of which is the source of opium.

positron, n. A particle first found in cosmic rays that has the same mass and magnitude as the electron, but is of the opposite charge.

pot, n. (See: marijuana.)

predator, n. An animal that preys on other animals.

primate, n. A mammalian class considered to be of the highest type, including man and the apes.

prism, n. In optics, a right cylindrical glass solid whose section is triangular and that readily disperses white light into the color spectrum.

progesterone, n. A female sex hormone that prepares the lining of the uterus for the development of an embryo.

prostaglandins, n. A group of acids found in high amounts in semen and whose properties and functions are as yet under study.

prostate gland. In men, a walnut-sized gland near the bladder, which secretes the carrier fluid for semen during male orgasm and which increases in size for the duration of a man's life.

prosthesis, n. The addition to the human body of an artificial part.

prostration, n. A state of complete exhaustion.

protein, n. A fundamental constituent of all living cells, formed by any one of several naturally occurring combinations of amino acids.

proton, n. The positively charged nucleus of a hydrogen atom.

protoplasm, n. Usually, a viscous, translucent material, which forms the physical environment of all living cells.

protozoan, n. A single-celled animal, which reproduces by biological fission.

psoriasis, n. A chronic skin condition marked by red patches with white scales.

psychoanalysis, n. A therapeutic approach to mental problems, founded on the theory that repressed but persistent desires are often causative factors.

psychosis, n. Serious mental disease or derangement.

psychosomatic, adj. Pertaining to a bodily manifestation of a mental problem.

puberty, n. The time at which sexual maturity is reached.

pulsar, n. At present, thought to be a rapidly spinning, collapsed star, from which varying radio signals are being emitted.

pupa, n. A quiescent stage of insect development between the larva and the adult stages; a cocoon.

purulence, n. Pus.

pus, n. A yellowish-white, opaque, creamy substance consisting of bacteria, white blood cells and dead tissue.

putrefaction, n. The decomposition of organic matter, often accompanied by foul smelling, incompletely oxidized products.

Q

Quaalude, n. A trade name for methaqualone.

quantum mechanics. A branch of physics that deals with atomic and nuclear phenomena and that assumes wave properties of particles and discreteness of energy states.

quark, n. A type of subatomic particle thought to be the building block of all subatomic particles.

quasar, n. A relatively small astronomical body of an unresolved nature, which may be extensively distant and moving at an exceptionally high speed.

quinine, n. An extract from cinchona bark used widely as a tonic to alleviate the symptoms of malaria.

R

race, n. Usually, a division of mankind characterized by skin color, all definitions of which have failed to be comprehensive.

radar, n. A scanning-beam radio device used for the detection of objects like airplanes and phenomena like thunderstorms.

radioactive, adj. Having the property of radioactivity.

radioactivity, n. The property of certain elements, like radium, uranium, and thorium, to emit rays of particles spontaneously from the disintegration of their nuclei.

radiocarbon dating. A method of age determination, or dating, which depends on the decay to nitrogen of a form of radioactive carbon.

radio wave, n. An electromagnetic wave used for sound transmission.

radium, n. A highly radioactive element found in minute quantities in pitchblende and uranium minerals.

recombinant DNA. DNA used to produce a new arrangement of genetic material.

red blood cell. A blood cell that carries oxygen from the lungs to the tissues and carbon dioxide from the tissues to the lungs.

red shift. The displacement toward redness in the light spectrum of a celestial object, which corresponds to an increase in wave length, due, for example, to the motion of a luminous object away from the observer.

refraction, n. The bending of a wave as it passes obliquely from one medium to another.

regeneration, n. Complete reformation and regrowth.

relativity, n. A branch of physics consisting of two parts, the special and the general theories, in which assumptions about the speed of light play a basic role and in which rest, or atomic, energy is a consequence.

REM sleep (rapid eye movement sleep). A stage of sleep characterized by rapid breathing and heart rate, and by bursts of rapid eye movements.

reptile, n. An animal, or its extinct ancestor, whose body has scales or bony plates and that crawls or moves on its belly or on small short legs, including snakes, lizards, turtles, and crocodiles.

resonance, n. The phenomenon of responding with maximum amplitude when the frequency of an applied force is the same as that of a given body.

respiration, n. The process of breathing.

retina, n. The back wall of the eye, which, like a screen, receives the image of an observed object from the lens and transmits it to the brain via the optic nerve.

rheology, n. The study of the deformation and flow of matter.

rheumatic fever. An acute disease of children and young adults, characterized by fever, pain in the joints, and inflammation of various heart tissues.

rheumatoid arthritis. A group of diseases characterized by pain, inflammation of the joints, and, possibly, the development of physical deformities.

Rh factor (Rhesus factor). A substance present in the red blood cells of most individuals, the absence of which can cause serious problems in infants and during transfusions.

riboflavin, n. Vitamin B_2.

ribonucleic acid (RNA). An acid carrier of the DNA genetic code away from the nucleus.

ribosomes, n. Small granules in the cell where proteins are synthesized.

rickets, n. A vitamin D deficiency disease of childhood, characterized by the bending of growing bones.

RNA (See: ribonucleic acid.)

Roentgen ray. An X ray.

rubella, n. German measles.

S

saccharin, n. A coal-tar crystalline product used as a sugar substitute.

sarcoma, n. A particular form of cancer occurring usually in bone cartilage or lymph tissues.

satellite, n. A small celestial object rotating around a larger one.

scarlet fever. An acute, contagious, bacterial disease, usually of childhood, characterized primarily by fever and a scarlet rash.

schizophrenia, n. A complex psychotic condition characterized by deteriorating ability to function in everyday life and some combination of the following: reduced tolerance for the stress of interpersonal relationships, loss of contact with one's environment, retreat into fantasy life often involving delusions and hallucinations, disturbed thinking involving distortion of common logical relations, movement disorder, and the hearing of voices.

Scopes trial. The trial of John T. Scopes in Dayton, Tennessee, in 1925, for violating state law by teaching the theory of evolution in a public school; also called the Monkey Trial.

scurvy, n. A vitamin C deficiency disease characterized by anemia, livid spots, and bleeding gums.

sea anemone. A lower form animal that lives in water and has the appearance of a plant.

sea urchin. A marine invertebrate with a round body and a radial arrangement of organs.

Second Law of Thermodynamics. A physical law that states that in every process involving heat energy, some of the heat is lost and cannot be recaptured.

seismic, adj. Pertaining to wave motion through the earth.

seismology, n. The study of earthquakes and related phenomena.

semiconductor, n. A crystallized substance, which is an insulator at very low temperatures and conducts electricity at an intermediate level at room temperature.

serous, adj. Pertaining to serum.

serum, n. A watery type fluid containing disease-specific immune bodies.

shoal, n. Usually, a sand bar or a reef.

shock, n. Biologically, an acute, circulatory failure marked by weak, rapid pulse, decrease in blood pressure, cold sweat, rapid breathing, dilated pupils, and dry mouth.

sickle cell disease. A disease caused by abnormal hemoglobin, which causes the red blood cells to assume crescent shapes and deprives them of the ability to carry oxygen to the tissues.

sidereal, adj. Pertaining to the stars.

silica, n. Any compound of the elements silicon and oxygen, which is the main component of more than 95 percent of the known rocks.

silicon, n. A natural element that appears abundantly in the earth's crust.

silicosis, n. A disease of the lungs caused by the inhalation of silicate dust.

slime mold. A primitive organism resembling both one-celled animals and fungi, found on damp earth and decaying vegetable matter.

smallpox, n. A serious, contagious disease marked by a pustular skin eruption.

smog, n. A mixture of smoke and fog.

solar corona. A luminous envelope surrounding the Sun.

solar energy. Energy derived directly from the Sun's rays.

solar flare. A sudden, intense brightening of the gaseous area round the Sun, which emits radiation.

solar plexus. A nervous plexus situated behind the stomach and in front of the aorta; the pit of the stomach.

solar system. The Sun and all celestial bodies that rotate around it.

solar wind. A massive flow of electrical particles away from the Sun.

solid-state electronics. Electronics without the use of vacuum tubes, relying chiefly on the use of transistors.

somatic cells. All the cells of the human body except germ cells.

sonar, n. An underwater navigational and detectional device, which uses echo sounding.

sonic boom. A cracking sound due to a shock wave in the air, caused frequently by an airplane traveling faster than the speed of sound.

space-time. A concept of relativity that unifies the concepts of space and time and utilizes the same units of measurement for each.

special relativity. The branch of relativity that assumes that the distance between two points is the straight-line distance.

species, n. A plant or animal group that can interbreed.

spectroscopy, n. The study of the nature of atoms and molecules from the light spectra that they emit.

speculum, n. An instrument for dilating body passages for viewing.

spermatozoon, n. A male sexual cell whose function is fertilization of an egg.

sperm whale. A large whale whose head has a closed, large cavity, which contains sperm oil.

sphincter, n. A ringlike muscle that is able to close a natural body opening.

spiral galaxy. Any galaxy configuration in which the stars seem to be arranged in a set of streamers, each of which is partially spiraled around the core.

spore, n. A primitive, unicellular, reproductive body produced, usually, by plants.

Sputnik, n. The name given to the first artificial earth satellite, launched by the USSR in 1957.

staphylococcus, n. A bacterial family that often seriously affects the skin and mucous membranes.

stem cell. A primitive cell that has not matured into a cell with a particular biological function. For example, a stem cell that is encoded to become a muscle cell does in time develop into a muscle cell.

stenosis, n. A narrowing of a passage opening.

Stone Age. The first known period of human culture in which stone tools were used.

Stonehenge, n. An ancient circular array of ditches and standing stones, found in England, and thought to be associated with religious and astronomical activities.

stratosphere, n. The upper portion of the atmosphere, usually above seven miles.

streptococcus, n. A bacterial family that often seriously affects the tonsils, throat, and lungs.

streptomycin, n. An antibacterial drug used in a variety of bacterial-caused diseases.

stroke, n. Cerebral bleeding or substantial reduction of blood flow to a portion of the brain, often resulting in paralysis, speech difficulty, loss of muscular coordination, or death.

strong force. A short-range force that binds together protons and neutrons in an atomic nucleus.

strychnine, n. A poisonous alkaloid.

sulfa drug. An antibacterial family of drugs used in a variety of bacterial-caused diseases.

sunspot, n. A dark spot that appears on the surface of the Sun and is associated with magnetic storms in Earth's atmosphere.

superconductor, n. A solid which, when cooled, suddenly loses all resistance to the flow of electric current.

supernovae, n. A star that suddenly increases greatly in luminosity and then changes radically from its original state.

suppuration, n. The production of pus.

suture, n. Medically, a stitch.

synapse, n. The point at which an electrical impulse passes from one nerve cell to another.

synchrotron, n. A special type of accelerator for charged particles, like electrons.

syndrome, n. Medically, the group of symptoms characteristic of a particular disease.

T

tectonic, n. Pertaining to structures resulting from deformation of the earth's crust.

tektite, n. A glassy, minerallike body of rounded but indefinite shape.

telemetry, n. The measurement of distant quantities by means of specialized electrical devices.

tensile strength. The smallest force necessary to pull along the length of an object in order to break it.

testosterone, n. A hormone of the testes responsible for the development of male sex organs and masculine characteristics.

tetanus, n. A painful, often fatal disease, marked by muscle spasms and caused by a microorganism that enters the body through a wound; lockjaw.

thermal reactor. A nuclear reactor that uses slow neutrons and generates large amounts of heat.

thermocline, n. Any curved line on a weather map, along which the temperature has the same value.

thermodynamics, n. The study of heat relationships and mechanics.

thermonuclear bomb, n. (See: hydrogen bomb.)

thermonuclear fusion. (See: fusion.)

thorax, n. That portion of the human body usually called the chest.

thrombophlebitis, n. The presence of a clot in a vein, often in a leg.

thrombosis, n. Usually, the formation of a clot in the blood or the lymphatic system.

thrombus, n. Usually, the clot of formation of thrombosis.

thyroid, n. A ductless gland located near the voice box whose hormone, thyroxin, has a profound influence on growth.
tic, n. A local muscle twitch or spasm.
tick, n. A blood-sucking parasite, slightly larger than a mite.
toadstool, n. Usually, an umbrella-shaped, poisonous fungus.
tonsils, n. A pair of prominent masses of lymphoidal tissue, one on each side of the throat, at the back of the mouth.
toxemia, n. Blood poisoning.
toxic shock syndrome. A condition of prolonged shock accompanied by high fever, vomiting, and diarrhea, which can be fatal and has been linked to the use of a particular type of tampon.
toxin, n. Any poison secreted by an animal or a plant.
tracheotomy, n. A surgical opening into the windpipe to allow breathing, which is otherwise not possible.
trajectory, n. The path of an object in motion, especially that of a missile, a rocket, or an artificial satellite.
transistor, n. A small, highly efficient, solid-state device for amplifying, controlling, and generating electrical signals.
trauma, n. Any wound or injury, physical or psychological.
trichinosis, n. A disease of the intestines and the muscles, caused by a parasitic worm, which is often found in improperly cooked pork products.
tropism, n. An involuntary movement of an organism in response to a stimulus, like the turning of a plant towards light.
trypsin, n. One of the enzymes secreted by the pancreas that aids in the digestion of food.
tsunami, n. The correct name for a tidal wave.
tubercle, n. A small, knotlike prominence; a nodule; in medicine, the specific lesion of tuberculosis.
tuberculosis, n. An infectious, bacterial disease characterized by the production of tubercles, especially in the lungs (pulmonary tuberculosis, consumption).
tularemia, n. A disease of rodents and domestic animals, which can be transmitted to man by insects, especially flies.
tumescence, n. A state of being swollen.
tumor, n. An abnormal tissue mass without inflammation and having no physical function.

turbine, n. A rotary type of engine or generator in which a fluid flows across a vaned spindle, causing it to rotate.

typhoid fever. A bacterial disease introduced into the body by food or drink and characterized by fever, intestinal catarrh, an abdominal eruption, diarrhea, and stupor.

typhoon, n. A tropical cyclone of the Philippine and China areas.

typhus, n. A serious disease transmitted by body lice and characterized by fever, exhaustion, red spots on the skin, and possible mental derangement.

U

UFO. (See: unidentified flying object.)

ulcer, n. A sore on the surface of a membrane which discharges pus.

ultrasound, n. Sound that is above the range of human hearing.

ultraviolet light. Light whose wavelengths are above those of visible light but below those of X rays.

umbilical cord. In mammals, the cord from the placenta to the navel of a fetus, through which the fetus is nourished.

uncertainty principle. A principle in quantum mechanics that states that one cannot know, simultaneously, an object's changes in both speed and position.

unicellular, adj. One celled.

unidentified flying object (UFO). An aerial object or optical phenomenon not explainable to the observer.

universe, n. In astronomy, all celestial constituents.

uranium, n. A radioactive element that is relatively heavy and readily usable as a source of atomic energy.

urea, n. A major and nitrogenous component of urine.

uric acid. A relatively small, odorless, and tasteless component of urine.

urine, n. In mammals, the fluid excretion from the kidneys, which is eliminated from the bladder through the penis or the external opening of the vagina.

urology, n. The study of the diseases of the urinogenital organs.

uvula, n. The fleshy, conical body attached to the soft palate, hanging centrally above the back part of the tongue.

V

vaccine, n. In general, any substance for preventive inoculation of a disease.

vagina, n. In the female mammal, a canal from the uterus to the external genital opening.

Valium, n. The trade name of the tranquilizer diazepam.

Van Allen belts. Zones of charged particles, trapped by Earth's magnetic field, several hundred miles up in the atmosphere.

vascular, adj. Usually, pertaining to the blood vessels.

vasectomy, n. Usually, a surgical cutting of the excretory duct of the testicles.

vasoconstriction, n. A constriction, or tightening, of a blood vessel.

vein, n. A large blood vessel that has valves and carries blood back from the capillaries to the heart.

venereal disease. Any of several diseases transmitted primarily by sexual intercourse.

venous, adj. Pertaining to veins.

vertebrate, n. An animal with a back bone, or spinal column.

viable, adj. In mammals, capable of living outside the womb.

virology, n. The study of viruses.

virus, n. Any of a group of submicroscopic, ineffective agents, which must utilize the machinery of other living cells to achieve their own reproduction.

vitamin, n. Any of a number of nitrogenous food substances that are essential for good nutrition.

vivisection, n. Dissection of a living animal for scientific purposes.

voltaic cell. A device for producing electricity by chemical means; a battery.

W

water strider. A small, slender, dark insect often seen running over the surface of a pond or a stream.

water table. The top portion of a section of earth that is saturated with water.

weak interaction. The local force that is responsible for the beta decay of radioactive elements.

white blood cell. A blood cell that helps maintain bodily health by destroying bacteria and other invasive foreign bodies.

WHO. (See: World Health Organization.)

whooping crane. A white, North America crane named for the particular type of sound it makes.

wind tunnel. A device for producing an air stream on models of aircraft that simulates the stress undergone in actual flight.

World Health Organization (WHO). A division of the United Nations concerned with implementing and maintaining minimal health standards worldwide.

X

X ray. An electromagnetic wave, similar in nature to light, but of a much shorter wave length, so that it cannot be seen by the human eye.

Y

yellow fever. An acute, tropical and subtropical viral disease of man and monkeys transmitted by mosquitoes.

Z

Zeeman effect. The shift of energy levels of an atom due to a magnetic field.

zoology, n. The study of animal structure and function.

zygote, n. A fertilized egg cell.